ENVIRONMENTAL PHILOSOPHY: THE ART OF LIFE IN A WORLD OF LIMITS

ADVANCES IN SUSTAINABILITY AND ENVIRONMENTAL JUSTICE

Previously ADVANCES IN ECOPOLITICS
Series Editor: Liam Leonard

PUBLISHED UNDER SERIES TITLE **'ADVANCES IN ECOPOLITICS'**

The Transition to Sustainable Living: Advances in Ecopolitics Volume 4
Edited by Liam Leonard and John Barry

Global Ecological Politics: Advances in Ecopolitics Volume 5
Edited by Liam Leonard and John Barry

Sustainable Justice and the Community Volume 6
Edited by Liam Leonard and Paula Kenny

Community Campaigns for Sustainable Living: Health, Waste & Protest in Civil Society Volume 7
Edited by Liam Leonard

Sustainable Politics and the Crisis of the Peripheries: Ireland and Greece Volume 8
Edited by Liam Leonard & Iosif Botetzagias

Enterprising Communities: Grassroots Sustainability Innovations Volume 9
Edited by Anna Davies

Transnational Migration, Gender and Rights Volume 10
Volume Editor: Ragnhild Sollund
Series Editor: Liam Leonard

PUBLISHED UNDER SERIES TITLE **'ADVANCES IN SUSTAINABILITY AND ENVIRONMENTAL JUSTICE'**

International Business, Sustainability and Corporate Social Responsibility Volume 11
Edited by Maria Alejandra Gonzalez-Perez and Liam Leonard

Principles and Strategies to Balance Ethical, Social and Environmental Concerns with Corporate Requirements Volume 12
Edited by Liam Leonard and Maria Alejandra Gonzalez-Perez

ADVANCES IN SUSTAINABILITY AND ENVIRONMENTAL JUSTICE VOLUME 13

ENVIRONMENTAL PHILOSOPHY: THE ART OF LIFE IN A WORLD OF LIMITS

EDITED BY

VOLUME EDITOR

LIAM LEONARD
Independent Researcher and Academic Consultant, Orange County, CA, USA

CO-EDITORS

JOHN BARRY
Queens University Belfast, Northern Ireland

MARIUS DE GEUS
Leiden University Netherlands

PETER DORAN
Queens University Belfast, Northern Ireland

GRAHAM PARKES
University College Cork Ireland

Emerald

United Kingdom – North America – Japan
India – Malaysia – China

Emerald Group Publishing Limited
Howard House, Wagon Lane, Bingley BD16 1WA, UK

First edition 2013

British Library Cataloguing in Publication Data
A catalogue record for this book is available from the British Library

ISBN: 978-1-78350-136-6
ISSN: 2051-5030 (Series)

CONTENTS

LIST OF CONTRIBUTORS

John Barry	Queens University Belfast, Belfast, Northern Ireland
Geoff Berry	Phoenix Institute of Australia, Victoria, Australia
Marius de Geus	Political Science Institute Leiden, Leiden, The Netherlands
Jason Dockstader	University College Cork, Cork, Ireland
Peter Doran	Queens University Belfast, Belfast, Northern Ireland
Thomas Duddy	National University of Ireland, Galway, Ireland
Liam Leonard	Independent Researcher and Academic Consultant, Orange County, CA, USA
Cara Nine	University College, Cork, Ireland
Graham Parkes	University College Cork, Cork, Ireland

ACKNOWLEDGEMENTS

I would like to acknowledge the following for their support and encouragement during the development of this volume: John Barry, Peter Doran, Maruis de Geus, Graham Parkes and all of the contributors and their families, and Tom Moylan, Maria-Alejandra Gonzalez-Perez, Lisa Brady, Paula Kenny, Eileen Leonard, Shane Leonard and Caitríona Leonard. We also remember our fellow contributor, the late Thomas Duddy, who sadly passed away while we were making this book.

This book is dedicated to the writer, poet and spiritual philosopher Janice Harter, for all her love and support.

INTRODUCTION

As Series Editor of the Advances in Sustainability and Environmental Justice series with Emerald Publishing, I am pleased to introduce a collection on the engaging topic of environmental philosophy for our 13th volume. It is often said that one shouldn't work with their heroes, but in the case of contributors like series co-editor John Barry, Marius de Geus, Peter Doran and Graham Parkes, I find myself working with intellectuals whom I consider to be both heroes and friends. Those familiar with their work will know that these writers combine philosophy, politics, activism and agency whilst instilling their writing with the type of insight and clarity that is required if we are to change understandings of the serious problems facing our planet due to environmental degradation and climate change.

Over time, I have had the pleasure of reading the works of leading environmental philosophers and writers such as Rachel Carson, Arne Naess, James Lovelock, Peter Singer, Gary Snyder, David Forman, Murray Bookchin, Robyn Eckersley, Vandana Shiva or environmental utopians such as Ralph Waldo Emerson, John Muir and Henry David Thoreau. I have engaged with green politics, deep green politics, ecotopianism and environmentalism. In 2004, I founded the *Ecopolitics Online Journal*, one of the first international academic websites dedicated to the politics of the environment. I also established Greenhouse Press, an online publishing entity dedicated to publishing environmental research. Greenhouse Press has gone on to publish material from every continent and on a diverse range of subjects. In time, *Ecopolitics* became an internationally successful series with Emerald Publishing, now followed by the current Advances in Sustainability and Environmental Justice series.

This latest collection in that series brings together the leading environmental writers mentioned above, alongside new and upcoming researchers in the field. As such, Volume 13 of the series represents an excellent set of perspectives on the philosophies and issues surrounding one of the greatest subjects known to humankind; that of nature and the environment. In this era of global challenges such as climatic change and all of the problems which surround it, this collection serves as a timely and salient reminder that the 'the Art of Life in a World of Limits' is one worth considering, and even philosophising about.

Liam Leonard
Series Editor

MACINTYRE AND HAVEL AS GREEN POLITICAL THINKERS: UNREASONABLE THINKING IN TURBULENT TIMES

John Barry

ABSTRACT

Purpose – *This chapter explores the ideas of Alasdair MacIntyre and Vaclav Havel and what these two thinkers can contribute to green political theory.*

Design/methodology/approach – *This chapter includes examination of some of the key works of Havel and MacIntyre and analysis of these works from the point of view of green political theory.*

Findings – *The section 'Havel and the Imperative to "Live in Truth": Dissent and Green Politics' explores Havel's thought with a particular emphasis on his ethicised notion of political action and critique ('living in truth') and his focus on the centrality of dissent (both intellectually and in practice) as central to political critique and action. The section 'MacIntyre as a Green Thinker: Vulnerability in Political and Moral Theory' offers an overview of MacIntyre interpreted as a putative green*

Environmental Philosophy: The Art of Life in a World of Limits
Advances in Sustainability and Environmental Justice, Volume 13, 1–29

ISSN: 2051-5030/doi:10.1108/S2051-5030(2013)0000013005

thinker, with a particular emphasis on his ideas of dependence and vulnerability. The Conclusion attempts to draw some common themes together from both thinkers in terms of what they have to offer contemporary green political thought.

Research limitations/implications – *What is presented here is introductory, ground clearing and therefore necessarily suggestive (as well as under-developed). That is, it is the start of a new area of exploration rather than an analysis based on any exhaustive and comprehensive knowledge of both thinkers.*

Practical implications – *This chapter offers some initial lines of exploration for scholars interested in the overlap between green thinking and the work of Havel and MacIntyre.*

Originality/value – *This is the first exploration of the connections between the works of Havel and MacIntyre and green political theory.*

Keywords: Alastair MacIntyre; Vaclav Havel; vulnerability; green political theory; dependency

INTRODUCTION

The reaching out towards thinkers not normally considered as part of the green political tradition should not be assumed to be a sign of the lack of ideas or vitality within contemporary green political thought. On the contrary, contemporary green theoretical discourse is in a state of very robust health, connected of course to the interconnecting and deepening ecological, resource and socio-economic crises unfolding in our anthropocene times. Indeed, the fact that we are living within what Thomas Homer-Dixon calls 'synchronous failures' (Homer-Dixon, 2006) does offer grounds for casting green political thought as a form of 'thinking in turbulent times (Barry, 2012), a necessary element of which in my view does imply reaching back into the history of political theory (both western and non-western) to see what can be learned from previous thinkers who lived in unsettling times (here of course Hobbes, Locke, Rousseau, Mill and Marx spring immediately to mind). In this manner, recent 're-readings' of the canon from a green theory perspective (such as one finds in the work of, inter alia, Biro, 2005; Cannavò & Lane, in press; Meyer, 2001) is to be very much welcomed.

Other reactions to the crisis can be seen in the tendency for some to abandon any hope of change and transformation and to begin preparation for the inevitable collapse of contemporary carbon-fuelled capitalism and to make plans for who comes after this collapse. This type of thinking and (in)action can be seen in aspects of the 'peak oil' perspective (Gupta, 2009; Orlov, 2011) or more recent movements such as the 'Dark Mountain' project in the United Kingdom, founded on 'post-environmentalism' (Hine & Kingsnorth, 2010) or the critiques offered by thinkers such as John Gary, Derrick Jensen or James Howard-Kunster.

Another related reaction is the embracing of a romantic-cum-nostalgic return to the past (or a version of the past), something that has both been part of green political theory and also used by critics against it, since its modern inception. A final reaction or approach, and one pursed in this chapter, is the engagement with contemporary thinkers who are not normally viewed as 'green' can also be beneficial in developing green thinking in these troubling and uncertain times. Apart from seeking to find common ground and areas of difference, such an exercise is also motivated by an impulse for green theory to widen and deepen its pluralistic character.[1]

This chapter draws heavily on some of my recent work (Barry, 2012) in which I have discussed the ideas of Alasdair MacIntyre and Vaclav Havel and what these two thinkers can contribute to green political theory. Here, I wish to extend the discussion and further develop that conversation. First a qualifier. What is presented here is introductory, ground clearing and therefore necessarily suggestive (as well as extremely under-developed). That is, it is the start of a new area of exploration rather than an analysis based on any exhaustive and comprehensive knowledge of both thinkers. In this respect, it is green reading of both MacIntyre and Havel, a movement from green theory towards the latter rather than in the opposite direction. Scholars of both MacIntyre and Havel will no doubt find much to disagree with perhaps in what follows, based as this chapter is on both an incomplete review of the full scope of the work of each thinker, coupled with an at best partial or novice's understanding of the full complexity of their thought.[2]

The section 'Havel and the Imperative to "Live in Truth": Dissent and Green Politics' explores Havel's thought with a particular emphasis on his ethicised notion of political action and critique ('living in truth') and his focus on the centrality of dissent (both intellectually and in practice) as central to political critique and action. The section 'MacIntyre as a Green Thinker: Vulnerability in Political and Moral Theory' offers an overview of MacIntyre interpreted as a putative green thinker, with a particular emphasis on his ideas of dependence and vulnerability. The Conclusion attempts to draw some common themes together from both thinkers in terms of what they have to offer contemporary green political thought.

HAVEL AND THE IMPERATIVE TO 'LIVE IN TRUTH': DISSENT AND GREEN POLITICS

In his posthumously published working notes, the Czech writer, politician and dissident activist Jan Patočka (and Havel's mentor) succinctly defines what he calls 'radical over-civilisation' (Nadcivilisace in Czech) as the 'clumping of all forms of life under the perilous form of industrial productivity' (Patočka, 2007, p. 265). What he meant by this was that in modern industrial society all forms of life and of thought were becoming increasingly valued only in accordance with a scale of efficiency determined by the model of industrial productivity allied to corporate and state forms of bureaucratic standardisation, efficiency and instrumental value. Patočka's concern was that this principle of instrumental rationality had serious negative consequences for free and critical inquiry. That is, it threatened the very spirit of inquiry and philosophical introspection that he thought comprised the essence of Europe, its value and potential contribution to improving the world. In this – pointing out the negative consequences of a one-sided rationality and technological mastery of nature (and humans) – he was simply expressing a view long articulated by movements (such as the Green and earlier humanist Marxist) and schools of thinking such as critical theory, especially the Frankfurt school (Biro, 2011) and thinkers from Weber to Horkheimer, Adorno, Marcuse and Habermas.

The struggle against Soviet communism by political dissidents such as Patočka, Havel and Solzhenitsyn do contain important lessons for contemporary 'dissidents' (in which I would include green politics), who both stand to their cultures at a critical and oppositional angle, and who 'make their stand' on daring to state that the foundations upon which the entire edifice of modern western societies are based are crumbling or weaker than usually thought of. I would suggest that greens who point out the mythic, ideological character of economic growth, and propose preparing for a future beyond or below orthodox economic growth and so on, are engaged in similar political work and struggle to Soviet era dissidents. Now, least I be accused of an unseemly comparison, while the moral courage needed by the likes of Havel and Solzhenitsyn were immeasurably greater than the courage needed to criticise economic growth and contemporary consumer society, and without seeking to place 'growth dissidents' on the same moral plane as 'communist dissidents', nevertheless I do contend the comparison can be made. In both cases courage was and is required; both sets of dissidents faced and face opposition and ridicule from entrenched loci of power – by the state, concentrations of economic power, dominant forms of thinking, the prevailing culture and the mainstream media. While I am resolutely not saying that consumer capitalism is the same as Soviet communism, there are sufficient similarities to justify such a comparison – from a green point of view both exhibit similar forms of control in the maintenance and extension of their social orders and 'regimes of truth'. This comparison also emphasises that green politics differs from both old left and new right in being critical of both state-centric politics and market fundamentalism in championing community and civic society against both the state and the market. Thus, to describe greens as 'dissidents' is not to ascribe to them some heroic status to enable them to claim some reflected glory and enhanced moral capital from other, more readily recognisable dissident movements and individuals. Rather it is to examine a green political position through emphasising its dissenting and oppositional character.

Here, in part, this involves revisiting or reintegrating the countercultural origins of the modern green movement within its current political project. In fact, revisiting and reviving the countercultural dimension of green politics helps bring together the characterisation of greens as 'sustainability pioneers' (Barry, 2012), and identifying them as self-conscious dissidents. By equal measure, the heterodox economic tradition, in which I place the various forms of green political economy, can also be seen as a form of dissident in its resistance to the 'regime of truth' of the neo-classical economics hegemony, the opposing of which as Deb points out is, 'far from easy, and likely to invite scorn and ridicule' (Deb, 2009, p. 510). As Peter Hay argues, and in reference to the discussion about the ideological and mythic character of the economic growth, 'The belief in the inevitability and desirability of [economic] growth is so deeply entrenched that insistence on the centrality of the goal of a steady-state within the green critique will consign environmental thought, in the eyes of John and Janet Citizen, to the land of the lunatic fringe' (Hay, 2002, p. 282). As I understand it then, to be a dissident and 'dissent' is to stand *within* one's society and political structure but not be *of* that social and political order in the sense of being reconciled and content with it. It is to be in a semi-permanent state of opposition to the culture and society within which one was perhaps born and raised, or just happen to live within, and whose dominant values and objectives one now rejects or seeks to radically reform, based on an alternative or new set of values. *To be a green dissident then is to never be satisfied.* And, it has to be said, to be a green dissident is also to invite ridicule and dismissal in so standing against the prevailing institutionalised forms of power and its associated culture, narratives and myths.

'Dissident' is perhaps a better and more accurate term to apply to greens than 'revolutionary', since while both share an opposition to the prevailing social order, revolutionary is more totalising clearly more antagonistic rather than agonistic. It is problematic and perhaps unproductive to seek to establish clearly defined distinctions between 'revolution' and 'dissent' and better to work with a more blurred, dynamic and malleable relationship between such transformative imperatives, motivations or strategies. Needless to say, this conversation between revolution and dissent is one that deserves more space than I have time for here.

Dissidents, as I see them, seek to direct a self-transforming present in a more radical direction, whereas revolutionaries typically seek the complete destruction of the existing order and then the construction of a new one. Greens as dissidents therefore begin from an acceptance of the inevitability of key aspects of this transition – primarily around climate change and the end of the oil age – and thus (perhaps inevitably) see an answer to 'what is to be done?' in terms of managing and shaping that inevitable transition, rather than building/re-building Dissident also seems less extreme and dogmatic in its critique and its demands, than those who advocate full-blow revolution. While I have not time here to defend it in more detail, my view is that it would be odd for greens to be dogmatic revolutionaries animated by a sense of the hopelessness of working within and through contemporary institutions or that there was nothing worth preserving within and from the contemporary social order. Though of course many critics of greens present them as dogmatic, ideologically motivated destroyers/threats to the current order (indeed Vaclav Klaus comes to mind here, as well as countless conservative and right-wing groups and thinkers), my own view is that the inherent pluralism and commitment to democracy and contingency within green thinking severely (but perhaps not completely) compromises this dogmatic and inflexible tendency. It is for this reason that green dissent could perhaps be (wrongly) described as somewhere on a continuum between 'reform' and 'revolution', a form of 'creative adaptive management' to create collective resilience in the face of actually existing unsustainability (Barry, 2012).[3]

In his essay 'The Power of the Powerless', Vaclav Havel uses the story of a greengrocer who unthinkingly displays his 'loyalty' to the regime by displaying a Communist Party slogan in his shop window. This the greengrocer does 'ritualistically, since this is the only way the regime is capable of acknowledging his display of loyalty' (Havel, 1978, p. 45). In a similar way, being a dutiful consumer and not questioning economic growth could also perhaps be regarded as the way in which loyalty to a dominant capitalist, consumer regime is ritualistically displayed, enacted and affirmed. It is for this reason, if not only this reason, that one completely misunderstands consumerism, consumption and being a 'consumer', if one views it solely individualistically as some economic-cum-metabolic act. As a public display of loyalty, consuming is first and foremost a collective act, an individual joining others in a shared activity and associated identity. So while critics such as Fromm are correct in highlighting the distinction in consumer culture between 'being' and 'having' (Fromm, 1976), what these analyses often miss is that consumption is also an act of 'belonging', form of affirmation of a particular sort of identity or subjectivity (Jackson, 2009; Keat, 1994). It is for this reason that a refusal to consume is so damaging to the modern political and economic order and why to consciously choose not to consume is perhaps one of the most politically significant acts one can do in a consumer society. And one that, the continual performance (or rather non-performance) of which, further marks one out as a dissident, part of 'the great refusal' to use Marcuse's term (Marcuse, 1964). That is, to question economic growth under consumer-capitalism is to be 'disloyal' to the prevailing order, to dissent. This dissent for Havel, as it is for greens, an ethical as well as a political act, an assertion of a different set of values, worldviews and principles and related forms of human identity and interests, ultimately the assertion of a different form of society or conception of 'civilisation' (a common reference point in Havel's work).

While for Havel living in what he calls the 'post-totalitarian' communist regime is a 'living a lie', I do not want to go so far and say that life in contemporary consumer capitalist democracies is in the same way to 'live a lie'. Rather what I would like to dwell upon is Havel's notion of 'living within the truth' and what this can offer for green dissidents. For Havel 'living within the truth ... can be any means by which a person or group revolts against manipulation: anything from a letter by intellectuals to a workers' strike, from a rock concert to a student demonstration, from refusing to vote in the farcical elections, to making an open speech at some official congress, or even a hunger strike' (Havel, 1986, pp. 59–60). Though clearly written with the then communist regime in mind, Havel's call to 'live in truth' is equally pertinent to consumer capitalism. As he puts it:

> The profound crisis of human identity brought on by living within a lie, a crisis which in turn makes such a life possible, certainly possesses a moral dimension as well; it appears, among other things, as a deep moral crisis in society. *A person who has been seduced by the consumer value system, whose identity is dissolved in an amalgam of the accoutrements of mass civilization, and who has not roots in the order of being, no sense of responsibility for anything higher than his or her own personal survival, is a demoralized person*. The system depends on this demoralization, deepens it, is in fact a projection of it into society. (Havel, 1978, p. 62; emphasis added)

Silence is of course a consequence and precondition for this demoralisation, and what power requires under consumer capitalism is passive and silent acquiescent as much as active participation. For Havel the re-appropriation of individual responsibility is something to be actively striven for. This reverses or balances the usual focus on rights and freedoms with which often 'progressive' critiques of consumerism are couched. In Havel's response to what Tim Jackson amongst others has called 'The Age of Irresponsibility' (Jackson, 2009). As Jackson notes, 'the "age of irresponsibility" is not about casual oversight or individual greed. The economic crisis is not a consequence of isolated malpractice in selected parts of the banking sector. *If there has been irresponsibility, it has been much more systemic, sanctioned from the top, and with one clear aim in mind: the continuation and protection of economic growth*' (Jackson, 2009, p. 26; emphasis added).

The struggle Havel describes from the 1968 'Prague spring' between 'the system' and 'the aims of life' (Havel, 1978, p. 66) resonate green concerns of the degradation of natural life-supporting systems and the undermining of conditions promoting human conviviality, quality of life and well-being (Barry, 2009; De Geus, 2009; Jackson, 2009). What Havel goes on to say about political change and strategy in the context of a consumer culture is pertinent and important for those seeking a transition away from unsustainability, 'Society is not sharply polarised on the level of actual political power, but … *the fundamental lines of conflict run right through each person*' (Havel, 1978, p. 91; emphasis added). This is a profound point, namely that it is difficult, if not impossible, to simply analyse actually existing unsustainability as an oppressive totalitarian regime in which there is an identifiable 'them' dominating 'us'. Under consumer capitalism, debt-based consumption etc. we who live in these societies are all implicated in its continuation. And while of course there are identifiable groups and institutions (such as large corporations, financial/wealth management firms, the leadership of mainstream political parties, key agencies of the nation-state such as Departments of Finance, global financial institutions such as the World Bank and the IMF, and what Sklair has called the 'transnational capitalist class') who do benefit more from actually existing unsustainability, we have to face up to the fact that 'ordinary people', that is everyone also contributes (unequally of course) to the 'mundane' operation of global capitalism and the exploitation of people and planet. The recognition of this is but another way of drawing attention to the fact that capitalism, the commonsense of neo-classical economics, etc. have achieved 'full spectrum' domination of hearts and minds, such that capitalism, and realistic critiques of it, need to be viewed as cultural (and indeed psychological) projects. It is for this reason that the Transition movement is interesting, since it adopts an explicitly cultural and psychological approach. Of course such cultural and psychological critical analyses are not exhausted by this movement and these cannot be a substitute for oppositional political struggle.

This 'cultural turn' in green politics is, to my mind, premised firmly on a notion of human flourishing that lies beyond production, 'supply side' solutions, 'competiveness' and increasing 'labour productivity'. This notion of flourishing is not anti-materialist. Let me make that abundantly clear, it is not an ascetic renunciation of materialism for its own sake, as if material life is intrinsically unworthy or does express valued modes of human being. Thus, I do not accept the Fromm-inspired view that materialism or indeed material consumption is simply a mode of 'having' and not 'being'. After all, the critique should be directed at consumerism and overconsumption, not materialism or consumption per se.

At a basic level one can see how communism and consumerism are two 'regimes of truth' imposing their version of the truth, exacting payment, compliance and subjectivity from their client populations, quelling, distracting and undermining dissidents, and using different but also some shared techniques to continue. And the appropriate dissident, progressive attitude and strategy against both is, for Havel, ultimately an ethical one, an ethical and political life-affirming 'reconstitution of society' (Havel, 1978, p. 115). That Havel conceives consumer-capitalist and communist societies as comparable can be seen in his view that:

> traditional parliamentary democracies can offer no fundamental opposition to the autonomism of technological civilisation, and the industrial-consumer society, for they, too, are being dragged helplessly along by it. People are manipulated in ways that are infinitely more subtle and refined than the brutal methods used in the post-totalitarian societies… *the omnipresent dictatorship of consumption, production, advertising,*

> *commerce, consumer culture, and all that flood of information.* (Havel, 1978, p. 116; emphasis added)

Some of the republican elements expressed in Havel's thought centre around 'responsibility' (Havel, 1986, p. 104). He maintains that the abdication of responsibility in the name of consumer choice – the reduction of political liberty to a consumer 'freedom of choice' – weakens the ethical and political capacities of citizens within liberal democracies. Liberal consumer–citizens then become 'victims of the same autonomism, and are incapable of *transcending concerns about their own personal survival to become proud and responsible members of the polis, making a genuine contribution to the creation of its destiny*' (Havel, 1978, p. 116; emphasis added). In this, Havel is articulating concerns very close to the type of green republicanism I have outlined elsewhere (Barry, 2012).

His concluding comments in 'The Power of the Powerless' also offer suggestive lines for interpreting the Transition movement. In a passage focusing the contours of what Havel calls the 'existential revolution' that is needed to re-new the relationship of humans to the 'human order and cosmopolitan responsibility', Havel notes that the structures needed to make this happen 'should naturally arise from below as a consequence of authentic "self-organization"; they should derive energy from a living dialogue with the genuine needs from which they arise, and when these needs are gone, the structures should also disappear ... The decisive criterion of this "self-constitution" should be the structure's actual significance and not just a mere abstract norm' (Havel, 1978, p. 119). A better description of the Transition movement's aims, motivations and objectives would be hard to find. Havel goes on to describe these new, provisional and practical structures 'post-democratic'. He describes the outlines of these 'authentic' political structures in this manner:

> Do not these groups emerge, live and disappear under pressure from concrete and authentic needs, unburdened by the ballast of hollow traditions? Is not their attempt to create an articulate form of 'living within the truth' and to renew the feeling of higher responsibility in an apathetic society really a sign of some rudimentary moral reconstitution? In other words, are not these informed, non-bureaucratic dynamic and open communities that comprise the 'parallel polis' a kind of rudimentary prefiguration, a symbolic model of those more meaningful 'post-democratic' political structures that might become the foundation of a better society? (Havel, 1978, pp. 120–121)

Fundamental here, I think, is Havel's call to responsibility and struggle against the prevailing political order when it undermines quality of life, perpetuates injustice or the denial or compromising of democratic norms.

In a similar vein Carla Emery puts it eloquently, 'People have to choose what they're going to struggle for. Life is always a struggle, whether or not you're struggling for anything worthwhile, so it might as well be for something worthwhile' (in Astyk, 2008, p. 204). Or to phrase it differently: get busy living or get busy dying.

MACINTYRE AS A GREEN THINKER: VULNERABILITY IN POLITICAL AND MORAL THEORY

With the exception of green political theory and feminist and ecofeminist thinking, issues of human (and nonhuman) vulnerability and associated ideas of dependence do not loom large in the canons of contemporary western moral or political thinking.[4] As Alastair MacIntyre has put it "an acknowledgement of anything like the full extent of that dependence and of the ways in which it stems from our vulnerability and our afflictions in generally absent [from western thinking]" (1999, p. 3).[5] Indeed much like the treatment of animals or the natural world within most books on contemporary political theory, vulnerability is either absent, warrants only the most brief of discussions (and then usually in terms of the 'standard special cases' or aberrations from the 'norm' of moral and political independence or autonomy); or comes last in the discussion.[6] MacIntyre has not only acknowledged vulnerability and dependency as important moral and political issues, but has gone further and made the argument that vulnerability and dependency are constitutive of the human condition, the non-acknowledgement of which compromises the practices and associated collective forms of life required for human individual flourishing and the realisation of common goods (MacIntyre, 1999, pp. 63–81).

It is surprising, given MacIntyre's acknowledgement of human vulnerability and dependence, and his more recent embracing of an understanding of the animal natures of humans – including the recognition of the affinity between humans as 'practical reasoners', and non-human species such as dolphins and great apes – not to mention his critique of capitalism, economic growth and embracing of a localist political perspective, that up to now there has not been more engagement with or use of his insights within green political theory.[7] Indeed, his book *Dependent Rational Animals* (MacIntyre, 1999) can be viewed as outlining a 'green' perspective, or at least a vision of moral and political life which is compatible with the main contours of green political theory. He begins with an all too rare acknowledgement of intellectual error. As he puts it in the preface, he admits that he,

> was in error in supposing an ethics independent of biology to be possible ... and this for two distinct, but related reasons. The first is that no account of the goods, rules and virtues that are definitive of our moral life can be adequate that does not explain ... how that form of life is possible for beings who are biologically constituted as we are, by providing us with an account of our development towards and into that form of life. That development has as its starting point our initial animal condition. Secondly, a failure to understand that condition and the light thrown upon it by a comparison between humans and members of other intelligent animal species will obscure crucial features of that development. *One such failure, of immense importance on its own account, is the nature and extent of human vulnerability and disability.* (MacIntyre, 1999, p. x; emphasis added)

In pointing out these two salient facts – that humans are not just *like* animals, but *are* animals, and also the ineliminable vulnerability and dependence of human beings – MacIntyre articulates arguments central to green normative thinking. Green thinking has as a central orientating theme the need to transcend the species barrier in our moral and political thinking. It also holds that humans are dependent in all sorts of ways (on fellow humans and the nonhuman world), and that this dependence is normatively significant, as well as politically salient. Against dominant strands of political thought and prevailing European culture, an acceptance of human vulnerability and dependence is viewed not as some 'abnormal' or deviation from the moral or political norm we should expect or strive towards for human beings, but as fundamentally constitutive of what it means to be human.[8]

The dominance of the self-sufficient, independent human being as the 'norm' of the 'free and rational person', which has been at the centre of western moral and political theorising, has of course already been challenged, notably by feminists, who have pointed out the existence of unequal and unjust gender and power relations which permits this fiction of the free and rational independent 'man' to exist. It is only under conditions where bodily, corporeal and basic human needs have been met, by others, that modern Enlightenment man can emerge into the public sphere, as a free, autonomous individual, safe in the knowledge that the 'messiness' of reproductive and domestic life and labour, the life and labour associated with children, birth, food, sex, home-making, etc. has been 'taken care of' by others, safely (though always never completely) sequestered away.

As Ariel Salleh (1999) and other eco-feminists such as Mary Mellor (Mellor, 1997) and Sherilyn MacGregor (2006), have pointed out, the 'embodied debt' (Salleh, 2009, p. 5), gendered caring/care work, and other unpaid labour necessary for the functioning of both the formal capitalist economy, and the liberal democratic political system (including liberal political theory) are occluded and missing. If they are acknowledged they are generally viewed as marginal within mainstream political discourse and discussion, and not central to its concerns, and often viewed as 'pre-political'. By 'embodied debt' Salleh means 'the debt owed North and South to unpaid reproductive workers who provide use values and regenerate the conditions of production, including the future labour force of capitalism' (2009, p. 5). Like other aspects of modern life, such as death, aging, dependency on nature, gendered labour is another set of relations and realities that are often conveniently 'sequestered' away from public inspection and debate.

MacIntyre argues that dominant paradigms of modern thinking are characterised by the 'refusal to acknowledge adequately the bodily dimensions of our existence. This failure or refusal is perhaps rooted in, and certainly reinforced by the extent to which we conceive of ourselves and imagine ourselves as other than animal, as exempt from the hazardous condition of "mere" animality' (1999, p. 5). In so acknowledging and recognising the moral import of the corporeality of human existence and the human condition, MacIntyre is in the company of theorists such as Ted Benton who remind us of the fact that, 'Humans are necessarily *embodied* and also doubly, ecologically and socially *embedded*, and these aspects of their being are indissolubly bound up with their sense of self and with their capacity for the pursuit of the good for themselves' (1993, p. 103; emphasis in original). And this also connects with long-standing ecofeminist arguments concerning the unrecognised and unpaid gendered labour and exploitation (Bennholdt-Thomsen & Mies, 1999) required to meet those embodied and embedded needs. Being so constituted not only is a 'brute fact' of the human condition but means that, as both MacIntyre and green thinking point out, constitute a range of limits (which are morally and politically significant) that need to be attended and comprehended for any realistic moral or political theory. We depend on other humans not simply for the meeting of biological and other species-specific and culturally specific needs, but also as MacInytre points out, to sustain us in our practical reasoning (MacIntyre, 1999, p. 96). It is as a species of 'reasoning animals', since this embodied account of humanity explicitly sees the continuity and connection between other animals and humans, that MacIntyre invites us to view ourselves.

It is as if dominant strands of western thinking conceive of our bodily and ecological existence as either a threat to some ideal or recently achieved notion of 'human subjecthood' (hence the knee-jerk and all too predictable reaction to quickly dismiss attempts to acknowledge and re-introduce bodily and ecological issues as signalling a regressive 'return to the past').[9] Or that talk of the body, vulnerability, reproduction and reproductive labour and our dependence on the non-human world, are essentially pre-political or non-political and resolutely 'non-ethical'. It is for this reason that the Greeks viewed the sphere of the home as one governed by violence and domination not freedom, and non-political rules which rendered the domestic sphere a sphere of 'idiocy' (Berry, 1989). For similar reasons political theorists such as Hannah Arendt are adamant that what she called the 'social' sphere be kept out of 'politics', least these 'pre-political' concerns 'contaminate' the proper and limited sphere of 'politics', understood as a sphere of autonomy, of 'action' not 'labour' (Arendt, 1958).

In this manner, to bring bodily concerns and those of our ecological embeddedness into the ethical-political realm is to commit a 'category mistake'. For example, like Arendt before him, Habermas is keen to establish clear boundaries between the 'political' and 'non-political', the 'ethical and non-ethical' or technological realms of human action, and to ensure that ethical concerns only relate to intra-human affairs. For Habermas, the only relation we have with the natural environment is an instrumental one, governed by productive, prudential and technical concerns about how best we can exploit it. His main concern, relating to what he sees as one of the dangers of modernity, is to prevent human social relations from being reduced to instrumental norms which are appropriate to the sphere of human technological manipulation of the natural world. That is, he does not want how humans treat and view each other to be within the same category of how we treat and view the nonhuman environment (Barry, 2007, pp. 92–116).[10]

A key issue to recall here is that the separation that vulnerability and dependency marks within humanity is not simply between different categories of humans – viz., the able-bodied and 'fully rational' as distinct from the disabled and those possessed of less than full rationality – the familiar 'special cases' of applied ethical and philosophical thinking. *But more importantly it delineates a division within the person.* That is, being vulnerable and dependent is not simply the possibility of belonging to a particular category of 'unfortunate' or 'defective' humans. Being vulnerable and dependent is something that happens to us all, to different degrees and at different stages and points in our lives. In other words, vulnerability and dependence are constant features of our lives, but having particular salience at certain times more than others (when we are young, sick, old or have children or older adults to care for). And they are constant and universal since they are constitutive and ineliminable aspects of what it means to be human. There is thus a biographical element to our inherent dependence and vulnerability – we simply cannot escape from it. As novelist and critic Susan Sontag puts it, 'Illness is the night-side of life, a more onerous citizenship. *Everyone who is born holds dual citizenship, in the kingdom of the well and in the kingdom of the sick*. Although we prefer to use only the good passport, sooner or later each of us is obliged, at least for a spell, to identify ourselves as citizens of that other place' (Sontag, 1978, p. 3; emphasis added). Illness here is of course indicative of an identity and experience of vulnerability and dependence and Sontag's important insights are discussed more fully below.

The fact that we are a species of primate (and all that implies in terms of instincts, evolutionarily elaborated forms of human unfolding – and flourishing – from infancy to adulthood and so on), is not a normatively unimportant issue in articulating and developing moral and political theories.[11] Not only is it essential to ground one's normative theory in an accurate rather than inaccurate account of what type of beings human beings are – to see ourselves as phenomenal rather than noumenal beings in the Kantian sense (we are material, corporeal, embodied beings not 'brains in vats'). But it is therefore also the case that any prescriptive theory – using the principle that 'ought' implies 'can' – needs to be based on an awareness of whether it is possible for beings so constituted as we are to live or act in the manner implied by the theory.[12] Finally, it is also significant that it is from having an accurate description or understanding of the type of beings we are, that we can have a plausible account of human flourishing, of what it means for human beings individuals and collectively to 'live well' (O'Neill, 1993). As Andrew Brennan suggest, 'in order to discover what sort of human life is valuable we must first consider what kind of a thing a human being is. Although there is, in my view, no complete answer to this question, we can … grasp one important aspect of human nature by reflecting on what are essentially ecological considerations' (1988, p. xii). And one would add biological and biographical considerations of vulnerability and dependence.

Returning to MacIntyre, he states that contemporary normative theorising, and everyday discourse and thinking are informed by dominant cultural narratives and ideas that actively encourage forgetfulness and/or minimisation of our biological and animal natures. As he puts it, 'there are too many contexts in which we allow ourselves to forget [our

Darwinian evolutionary history], a cultural tendency that is reinforced by too exclusive an attention to and exaggeration of what does indeed distinguish human beings from members of all other species' (1999, p. 12). One implication of this is the danger of morally objectionable forms of treatment and misrecognition of the natures of non-human animals. As animal rights and welfare theorists have pointed out, some of this misrecognition and associated ill-treatment of animals can be traced to an ignorance (wilful or other) of the similarities (and indeed solidarities) that do exist across the species boundary between humans and non-human animals (Clark, 1977; Midgley, 1995; Singer, 1990). Another implication of this forgetfulness is a fundamental misrecognition of the essentially dependent and vulnerable character of humans, and the contingency that surrounds them and which together constitute their circumstances of flourishing.[13] Here, one could point out that the 'arrogance of humanism' (Ehrenfeld, 1978) which marks much of western thinking and culture, is based on the strong sense of *invulnerability* that pervades Enlightenment anthropocentric thinking in general, and dominant strands of liberal thinking in particular.

It is also worth noting the connection between the eradication or downplaying of dependence and vulnerability and how these concepts of invulnerability, independence and individualism have become central to the dominant western linear view of 'progress'. In the 'western imaginary' (cultural and normative) vulnerability signifies not just something to be overcome, but also something which was dominant at an earlier stage of human evolution and history. It is an unwelcome 'reminder', representing an earlier stage of human evolution. It is in this respect that the Enlightenment vision of progress (itself based on an Augustinian notion of progress) is the heroic story of human reason overcoming the pre-modern human condition of ignorance, poverty, want and superstition. Echoing the premise of the founding of the welfare state and the slaying of the 'five giants' (squalor, disease, idleness, want and ignorance), *vulnerability is something to be eradicated, or sequestered, not a condition to be acknowledged.* In Susan Sontag's analysis of illness as a metaphor in modern society (Sontag, 1978), discussed further below, notions of vulnerability and dependency are akin to (curable) diseases, reminders of 'lower' and 'ruder' stages of social evolution (in the terms used by eighteenth century Scottish Enlightenment thinkers such as Adam Smith and Adam Ferguson), and therefore constitute things to be transcended and overcome.

There does seem to be some (albeit limited) acknowledgement of this blind spot in contemporary political theorising around human vulnerability.

One form it takes – which does not make the link to the essentially biological and animal character of humans as outlined above – is given by Anderson and Honneth (2005), who focus on the excessive attention given to individual independence as the proto-typical example of autonomy within liberal theory. As they put it,

> the idea that individuals realize their autonomy by gaining independence from their consociates... within culture at large, the images that accompany the emergence of this conception of autonomy suggest that *any* constraints reduce an individual's autonomy. As part of this development, however, an individualistic conception of personal autonomy has crept into modern theories of social justice. The point of creating a just society comes to be seen as allowing people to be as little dependent on others as possible. The conceptual consequences of this individualistic strain have been massive. *They include not only the idea, for example, that autonomy increases with wealth but also the idea that unchosen membership in a community represents a threat to personal autonomy* (Anderson and Honneth, 2005, p. 128; emphasis added).[14]

The connection they make here between 'autonomy' and 'wealth' is important and is linked to the ideological, mythic and normative support for 'economic growth' which operate alongside more structural drivers. While leaving that discussion until later, at this point, I would say that the point Anderson and Honneth make here about this moral dimension to economic growth and wealth creation under capitalism is one that others have made. Perhaps one of the most eloquent of these is Benjamin Friedman's *The Moral Consequences of Economic Growth* (Friedman, 2006), who presents a strong case for the positive moral benefits of orthodox economic growth. Indeed, he argues that the growing incivility and intolerance within America are a direct consequence of the decline in middle class living standards (*ibid.* pp. 5–6). If this were true this would represent a serious challenge to the argument I make later for a 'post-growth' society, and associated argument of less consumerist forms of human flourishing. However, I note in passing here that Friedman pays little attention to growing socio-economic inequality and the decline in community in recent decades, as perhaps being a greater cause of the growing political incivility, intolerance and declining openness and generosity within American society. Nor does he consider that perhaps what the rise in incivility and intolerance in America signals is that orthodox economic growth has passed a threshold where it is now 'uneconomic growth'.

Anderson and Honneth continue their analysis by suggesting that, 'If, by contrast, we recognize that individuals – including autonomous individuals – are much more vulnerable and needy than the liberal model has traditionally represented them as being, *a very different picture of the demands of social justice emerges*' (*ibid.*, p. 129; emphasis added). This is an extremely significant issue they raise. For them, 'one's autonomy is vulnerable to disruptions in one's relationship to others … agents' social vulnerability in the ways in which being able to lead one's own life is dependent upon one's being supported by relations of recognition' (*ibid.*, p. 130). This leads to more explicit integration of concerns about institutional support for respectful recognition (which can only arise in the context of social relations rather than individual self-reflection) as constitutive of the achievement of autonomy within liberal theories of justice. Thus, Anderson and Honneth go some way in recognising the inter/dependent social nature of human beings.

Unfortunately they do not examine potentially 'unhealthy' forms of recognition seeking, through 'status competition' for example, or excessively consumerist identities (Jackson, 2009; Kasser, 2008), or that there are healthy forms of human identity and flourishing which are possible beyond orthodox economic growth and consumerism. Equally unfortunately they do not, as is characteristic of most contemporary political theorising, directly relate this to our distinctly *ecological* dependencies. This neglect of ecological dependencies is a point also made by Dobson in his analysis of MacIntyre's focus on social as opposed to ecological dependencies (Dobson, 2009). Anderson and Honneth do not explore the possibilities for a theory of justice based on a full recognition of our multiple vulnerabilities and dependencies. As Fineman notes, 'Understanding the significance, universality, and constancy of vulnerability mandates that politics, ethics, and law be fashioned around a complete, comprehensive vision of the human experience if they are to meet the needs of real-life subjects. Currently, dominant political and legal theories are built around a universal human subject defined in the liberal tradition' (Fineman, 2008, p. 10), indicating that a 'post-liberal' political vision is required.

The acknowledgment of the human need for recognition as arising only in the context of social relations does have a biological and evolutionary basis. Humans are after all social beings (a species of social primate) and, as such, require not just to live in the *presence* of other humans. But also the *participation* of and *ongoing interaction* with other humans beings is required for developing central human capacities (such as language use or practical reasoning), but also as major contributors to (or negative influences on) human well-being and flourishing. Anderson and Honneth underestimate the significance for the development of a theory of justice of a full acceptance and acknowledgement of human vulnerability and dependence. A theory of justice which is based on something like MacIntyre's notion of 'rationally acknowledged dependence' (MacIntyre, 1999, pp. 119–129), as opposed to striving for (ir)rational (?) independence would be different, perhaps very different, from the dominant accounts of justice within contemporary (liberal) political theory.

It is beyond the present study to provide an answer to the question raised above in terms of what a full theory of injustice and associated account of social order would like if they fully acknowledge and embrace human vulnerability. MacIntyre suggests a starting point in holding that what is needed is to, 'envision ... *a form of political society in which it is taken for granted that disability and dependence on others are something that all of us experience*...and that consequently our interest in how the needs of the disabled are adequately voiced and met *is not a special interest ... but rather the interest of the whole political society*, an interest that is integral to their conception of the common good' (MacIntyre, 1999, p. 130; emphasis added). Apart from undermining the usual categorising of people with disabilities as 'special cases', and breaching the ethical boundary between 'human' and 'nonhuman', a theory of justice and society based on human vulnerability and dependence rather than invulnerability, self-sufficiency and independence moves us, I would suggest, in the direction of an (eco) feminist 'ethic of care' perspective (Macgregor, 2006; Mellor, 1997; Salleh, 2009), combined with an acknowledgement both of the value and ineliminable normative presence of nature's 'otherness' in our deliberations about justice (Baxter, 2005; Cripps, 2010; Nussbaum, 2004; Schlosberg, 2007; Smith, 2006, 2008). While I am now (in a way I previously was not) persuaded that it is possible to include human-nature exchanges within an account of justice, that it 'nature counts', I am less sure about how and in ways it does, and how its inclusion affects the architecture of justice and politics. However, I do think an account of justice and politics based on vulnerability and dependency would give us a more thickly relational, contextually embedded theory of justice, and a more embodied one too (Carolan, 2009; Gabrielson, 2008; Gabrielson & Parady, 2010). It would deliver an account of associated notions of human flourishing and the good life which are much more collective than those offered by dominant accounts of liberalism (or some existing accounts of green politics). At the very least, such an conception of justice and politics would also take a 'whole life cycle' view of the individual rather than, as most liberal accounts do, take one period of that life cycle (i.e. healthy adulthood) and makes this paradigmatic and representative of what it means to be a flourishing and 'normal' human subject.

Illustrative here is MacIntyre's use of Aristotle's discussion of the vice of the *megalopsychos* – an unwillingness to acknowledge what one has received on the grounds that this would compromise one's superior sense of self-sufficiency, independence, close to the sense of enlightenment/liberal 'autonomy' as outlined by Anderson and Honneth above. According to Aristotle, the *megalopsychos* 'is ashamed to receive benefits, because it is a mark of a superior to confer benefits, of an inferior to receive them' (*Nicomachean Ethics*, 1124b 9–10). In this manner Aristotle might have been bemused at the dominant liberal view that, to repeat Anderson and Honneth, '*The point of creating a just society comes to be seen as allowing people to be as little dependent on others as possible*' (2005, p. 128), when such dependence and interdependence, and the relations and practices that flow from them, are both constitutive not just of being human or the human condition, but indeed central features that need to be taken account of in achieving human flourishing. According to MacIntyre:

> So the *megalopsychos* is forgetful of what he has received, but remembers what he has given, and is not pleased to be reminded of the former, but hears the latter recalled with pleasure. *We recognize here an illusion of self-sufficiency, an illusion apparently shared by Aristotle, that is all too characteristic of the rich and powerful in many times and places*, an illusion that plays no small part in excluding them from certain types of communal relationship. For like the virtues of giving, those of receiving are needed in order to sustain just those types of communal relationships through which the exercise of those virtues first has to be learned. (MacIntyre, 1999, p. 27; emphasis added)[15]

The point here is not to somehow celebrate or exacerbate our multiple vulnerabilities and dependent natures. Or to eschew all modern efforts to minimise harmful vulnerabilities, in the name of some 'authenticity' linked to a disavowal of the multiple benefits of modernity and modern forms of life, technology and its other achievements. Thus, it is important here to state that an acknowledgement of vulnerability and dependence should not be read as endorsing all of MacIntyre's ideas. Such as, for example, his anti-industrial tendencies in gravitating towards parochial and distinctly 'pre-modern' forms of social life, and his predilection for Donegal farming cooperatives, 'traditional' and non-industrial forms of collective production as the only exemplars of meaningful work practices (Breen, 2007, p. 411). Even though there is much in MacIntyre's localist vision which does have merit and is compatible, in a reworked form I would suggest with key aspects of a green political vision. Neither should it be read as supporting that 'primitivist' strand of 'hard green' political theorising (Humphrey, 2008; Jensen, 2002, 2010; Somma, 2009).

In short, what greens and MacIntyre seem to be asking are the following questions:

1. What forms of social and economic life are possible which fully acknowledge human vulnerability and dependence, and which see these as constitutive of the human good (since they are constitutive of what it means to be human) and not as aberrations to be eradicated?
2. Are conceptions of humanity and the human good, and political theories based on them, which fail to acknowledge vulnerability and dependence, 'fit for purpose' as practical guides to action?

A normative or social-scientific study of the relationship between society and the natural environment requires that we address the following question: 'how do we open up to investigation the relationships between humans and the rest of nature, without letting in the "Trojan horse" of biological determinism?' (Benton & Redclift, 1994, p. 4). That is, how, on the one hand, are we to study the ecological and biological aspects of the relationship of the human species to its natural environment and human evolutionary biology? While avoiding, on the other hand, thinking about and studying humans as if they were simply another species, and thus explaining human behaviour using the same biological and ecological models and conceptual analysis used in the case of other species? It is quite obvious that humans are a particular species, living in particular environments like other species on the planet, and thus it seems inappropriate to reject a biological or ecological approach to the study of human society and behaviour. As Benton puts it,

> We can, and, I think, we should, continue to view humans as a species of living organism, comparable in many important respects with other social species, as bound together with those other species and their bio-physical conditions of existence in immensely complex webs of interdependence, and as united, also, by a common coevolutionary ancestry. To say this much *is* to be committed to a naturalistic approach, but not necessarily to a reductionist one. It is to be committed to recognizing the *relevance* of evolutionary theory, physiology, genetics and, especially ecology itself, as disciplines whose insights and findings are pertinent to our understanding of ourselves. (Benton, 1994, p. 40; emphasis in original)

Benton I think offers a response to Dobson's critique of MacIntyre, in acknowledging that we need to attend to both biological and ecological dimensions of human vulnerability and dependency as constitutive elements of what it means to be human. He argues that we should not confuse the *relevance* of the natural sciences in investigating social phenomena with these sciences offering a *full* explanation of those phenomena. This is the mistaken path taken by socio-biology and other deterministic accounts of 'reading off' and prescribing human behaviour from human evolutionary biology (the idea of the 'selfish gene' for example), or from the observed behaviour of other nonhuman animals (Barry, 2007). The point Benton makes, here following Marx to a large extent, is that we need to generate a conception of our species-specific forms of knowledge pertinent to identifying our species-specific modes of flourishing. Though sharing with other nonhuman species similar problems and similar characteristics, the biological need for food, for example, humans are *not* the same as other species. Our particular species-specific natures, needs and modes of flourishing are such that we are different from other species, not least in the (not unlimited) cultural variation of how we satisfy those needs and flourish. But being different does not mean that we as a species are somehow radically separate or 'superior' to other species. In this way, Benton, seeks to transcend the dichotomy between 'environment' and 'society', as an important step in developing a more ecologically sensitive form of social theorising, necessary for developing and thinking about vulnerability. And in this he offers an analysis of human-nonhuman dependency and continuity, which compliments that of MacIntyre and his focus on intra-human dependency and vulnerability. And both present challenges to the dominant Enlightenment notion of the post-natural, and radically independent and autonomous self.

CONCLUSION

So what can green political through learn from an engagement with the two thinkers briefly sketched above? Being short of time and space I will focus on three shared themes that are suggestive of further study and analysis where green political theory can benefit from a deeper appreciation of the idea of Havel and MacIntyre.

Both thinkers are critical of modern industrial civilisation (in both its capitalist and communist forms) and here there are definite echoes of a view of green politics as similarly being 'beyond left and right', of articulating a distinctive position for example that emphasises the social and community as the preferred socio-economic modality beyond and over both 'state' and 'market' (including viewing state and market where these institutions are warranted as in the service of community). Thus, both Havel and MacIntyre share a green political sensibility around prioritising community (though that term is, of course, radically under defined here nor is it unproblematic), both as a source of normative orientation and as a practical institutional mechanism for meeting human needs (material and non-material). In this they also join other 'proto-green' and radical thinkers such as Karl Polanyi and Ivan Illich, and while of course a huge issue, I think a non-romantic reassertion of community is something that both thinkers and greens share.

Both are courageous thinkers in self-consciously refusing to go with mainstream culture, political status quo, or dominant intellectual current. While one may of course disagree with their analysis and suggested solutions to the problems they identify (perhaps more so with MacIntyre than with Havel), they both express a strong sense of having the courage of their convictions and standing at odds – and perhaps alone or in a minority – (like green politics) against the prevailing order.[16] Thus, both are dissent thinkers – most clear in Havel, whose writing and political activism are testaments to his courageous dissident against a totalitarian regime, but also in MacIntyre in his sometimes awkward railing against modernity, liberalism and industrial capitalism.

The idea of the good of politics is I think clear in both thinkers and in particularly they both express or their ideas are not incompatible with a distinctly 'civic republican' approach to political theory, critique and action. As Havel once noted during his presidency, 'Despite all the political misery I am confronted with every day, it is still my profound conviction that the very essence of politics is not dirty' (Havel, 1991). Whether it's the sense of active citizenship that both see as a key element of a reformed democratic politics (including the extension of democratic norms and participation to economic organisation), the way both articulate a non-arrogant humanism that accepts and works with notions of vulnerability, provisionality and limits or the ethical and political implications of responsibility as a key organising concept, both Havel and MacIntyre could be classified as 'republicans'. Thus they are both interesting for those (such as myself) who articulate a republican version or reading of green politics (Barry, 2012; Cannavò, 2011) and for whom the issue of 'leadership' is a missing element of green democratic political theory and action.[17]

Like contemporary green dissidents, Havel and MacIntyre can be viewed as 'productively unreasonable' and awkward. What is meant by 'unreasonable' in this instance is expressed in George Bernard Shaw's comment that, 'The reasonable man adapts himself to the world. The unreasonable man persists in trying to adapt the world to himself. All progress, therefore, depends upon the unreasonable man'.[18] And if no other reason than further developing and strengthening this essential and much needed quality of unreasonableness within contemporary green political thinking and action that Havel and MacIntyre in very different ways have much to teach.

NOTES

1. For some this could also, of course, be construed by some as a form of opportunistic or inappropriate 'intellectual colonisation' or 'imperialism' on behalf of green political theory, with the current confluence of crises used to present green theory as *the* over-arching politics of the anthropocene, that is, for green political concerns to become the standards against which all politics must be judged. Such tendencies and motivations do, I think, exist in green thinking (including my own contributions over the years), but this is an issue not pursed in this chapter. Expressions of this can be found in the work of Baxter (1995).

2. For those interested in deeper analyses of both thinkers, see Pontuso (2004) and Knight (1998).

3. It is also worth remembering here in talking of dissident, that ecology itself was once viewed as 'the subversive science'. As Murdoch and Connell note, an ecologically informed analysis 'has to be subversive or the ecologist will become merely subservient' (Murdoch & Connell, 1971, p. 323). It can also be suggested that such undogmatic forms of thinking and acting can have negative consequences in leading to either incoherence and dyspraxia and/or ineffective forms of moralised critique that lack political traction.

4. Political and ethical theorists have examined the normative status and treatment of 'vulnerable' individuals and groups, usually the standard 'moral patients' of the young, future generations, non-human animals, those in a vegetative state or with diminished capacity for reasoning and action. A noticeable exception here is Robert Goodin's *Protecting the Vulnerable* (Goodin, 1985).

5. Prentice, from a psychological perspective put this denial of dependency even more starkly: 'Is the human species suicidal? Apparently so – engaging in behaviour that is destructive to everything on which it depends, but apparently in serious denial of this … Unresolved dependency needs? Absolutely! We act as though we are not totally dependent on these others, as though can afford to abuse everything … of which our world is made … We seem to have an overweening narcissism, such that all other species and elements of the world appear to be there to please and gratify our every whim' (Prentice, 2001, p. 7).

6. I cannot be the only person struck by the pattern one finds in reading political theory books in which words like 'environment/environmental', 'nature/natural world', 'animals' and 'nonhumanity', are either absent from the index, or invariably (if it's a textbook usually) placed at the end of the work in question. It is as if the priority accorded to these issues can be seen not only in the number of words devoted to discussing them but also in the place/order they are presented. While one cannot judge a book by its cover, its table of contents or index (as any hard-pressed academic or researcher knows!) can tell you a lot about it.

7. Exceptions include Dobson (2009) and Calder (2009).

8. Brennan makes a similar point, but sees this impulse towards denying our vulnerability and dependence or seeing it as something to be overcome and resented, not as something particular to western post-Enlightenment culture, but something that is at the centre of the human psyche. According to her, while discussing the meaning of the Biblical account of the origin of the Devil, 'He was driven by, very simply, by the desire to be boss. From this desire comes a fantasy from which all paranoia and its attendant miseries are born. In this fantasy, the state of being created by another, and dependant on that other, is interpreted as the state of being controlled. The other or others want to keep you in your place. They created you to feel superior. If not for their ignoble ambition, you could be, well, God. This fantasy founds the psyche. In the fullest sense of the word, it is foundational' (Brennan, 2000, p. 7). From this fantasy and impulse of control, mastery and domination come many demons from hubris, pride, to the arrogance of humanism.

9. Exceptions in the western philosophical tradition would include Merleau-Ponty, William James, John Dewey and the contemporary writer on somaesthetics, Richard Shusterman (Shusterman, 2008). I owe this point to Peter Doran.

10. This aversion to human vulnerability and dependence is particularly acute in liberal thinking (as well as in an exaggerated form in thinkers such as Nietzsche, Sorel and far-right libertarians such as Ayn Rand, all possessed in different ways of a 'heroic' conception of the 'free' human individual). The stress on independence and invulnerability within liberalism explains its focus on the individual, as well as the common critique of the ontological and epistemological atomism within dominant strands of liberal thinking. This liberal disposition may explain some of the extreme aversion to green thinking by thinkers such as John Gray and Anthony Giddens, based not simply on their suspicion that greens are authoritarian and therefore a threat to freedom and the individual, but also – especially in Giddens case – on the grounds that greens transgress the species boundary in a manner that liberal ordering principles cannot cope with.

11. See Benton (1993, 2002), Barry (2007, Chap. 10), Dryzek and Schlosberg (1995) and Baxter (2007), for analyses of the ethical and political implications of evolutionary biology. Others such as Mary Midgley (1994, 1995) and Stephen Clark (1977, 1982, 1999) have done much over a long time to demonstrate the significance of biology, ecology and evolution for moral theorising.

12. This is another dimension of the 'applied' approach to political theory outlined in the introduction.

13. Dobson makes a valid point in his discussion of MacIntyre, noting that 'MacIntyre's understanding of "dependence" falls short of what is required for a full understanding of our condition as biological (better, "ecological") beings' (Dobson, 2009, p. 158), on the grounds that he foregrounds social over ecological relations of dependency.

14. Much has been made of the latter point in terms of the threat to (liberal) conceptions of autonomy from community membership in terms of debates between liberals and more communitarian orientated political theory, the connection between wealth accumulation and autonomy has typically been an object of critique by Marxist-inspired political theory and more latterly green political theory in offering a profound critique of the economic growth imperative at the heart not just of

capitalist economic logic, but also dominant strands of western political thinking such as liberalism (Barry, 2001).

15. A similar point is made by David Suzuki (1997) and connects with the 'socially embedded' character of the 'social economy' and its associated modes of provisioning as an 'economy of regard' as Offer puts it (Offer, 2006).

16. MacIntyre is or can be interpreted to be (at least according to many critics) 'anti-modern' in seeking at times to revert to local communities (and here of course there are links with some strands in the green thinking which also emphasise the local, the small-scale and are antithetical to bureaucracy, mass forms of production and consumption and generally seek (or predict) a radical 'de-complexification' of social, economic and political life (Sale, 1980; Hine & Kingsnorth, 2010).

17. Havel in particular is interesting for green theorists in relation to his ideas about the role and function of intellectuals in political struggle and political life.

18. And, least it is not immediately obvious, the unreasonableness and dissent expressed is one founded ultimately on a politics and ethics of hope, in part based the good (as well as necessity) of politics (and the responsible use of political power).

REFERENCES

Anderson, J., & Honneth, A. (2005). Autonomy, vulnerability, recognition and justice. In J. Christman & J. Anderson (Eds.), *Autonomy and the challenges to liberalism: New essays* (pp. 127–49). Cambridge: Cambridge University Press.

Arendt, H. (1958). *The human condition*. Chicago, IL: University of Chicago.

Astyk, S. (2008). *Depletion and abundance: life on the new home front*. Gabriola Island, Canada: New Society Publishers.

Barry, J. (2001). Greening liberal democracy: Theory, practice and political economy. In J. Barry & M. Wissenburg (Eds.), *Sustaining liberal democracy: Ecological challenges and opportunities* (pp. 59–81). London: Palgrave.

Barry, J. (2007). *Environment and social theory* (2nd ed.), London: Routledge.

Barry, J. (2009). Choose life not economic growth: Critical social theory for people, planet and flourishing in the age of nature. *Current Perspectives in Social Theory*, *26*, 93–113.

Barry, J. (2012). *The politics of actually existing unsustainability: Human flourishing in a climate-changed, carbon constrained world*, Oxford: Oxford University Press.

Baxter, B. (1995). *Must political theory now be green? Contemporary Political Studies 1*. Belfast: Political Studies Association.

Baxter, B. (2005). *A theory of ecological justice*. London: Routledge.

Baxter, B. (2007). *A Darwinian worldview: Sociobiology, environmental ethics and the work of Edward O. Wilson*. Aldershot: Ashgate.

Benton, T. (1993). *Natural relations: Ecology, animals and social justice*. London: Verso.

Benton, T. (1994). Biology and social theory in the environmental debate. In M. Redclift & T. Benton (Eds.), *Social theory and the global environment*. London: Routledge.

Benton, T., & Redclift, M. (1994). Introduction. In M. Redclift & T. Benton (Eds.), *Social theory and the global environment* (pp. 1–19). London: Routledge.

Berry, C. (1989). *The idea of a democratic community*. Hemel Hempstead: Wheatsheaf Books.

Biro, A. (2005). *Denaturalizing ecological politics; Alienation from nature from Rousseau to the Frankfurt school and beyond.* Toronto: University of Toronto Press.
Biro, A. (Ed.). (2011). *Critical ecologies: The Frankfurt School and contemporary environmental crises.* Toronto, ON: University of Toronto Press.
Breen, K. (2007). Work and emancipatory practice: Towards a recovery of human beings productive capacities. in *Res Publica, 13*, 381–414.
Brennan, A. (1988). *Thinking about nature: An investigation of nature, value and ecology.* London: Routledge.
Brennan, T. (2000). *Exhausting modernity: Grounds for a new economy.* London: Routeldge.
Calder, G. (2009). Listening, democracy and the environment, in-spire, 4:2. Retrieved from http://www.in-spire.org/archive/vol4-no2/Calder42.pdf. Accessed on February 8, 2011.
Cannavò, P. & Lane, J. (in press). *Greening the canon: Classic political theory and environmentalist questions.* Boston: MIT Press.
Carolan, M. (2009). I do therefore there is: Enlivening socio-environmental theory. *Environmental Politics, 18*(1), 1–17.
Cannavò, P. (2010). To the thousandth generation: Timelessness, Jeffersonian republicanism and environmentalism. *Environmental Politics, 19*(3), 356–373.
Clark, S. (1977). *The moral status of animals.* Oxford: Clarendon Press.
Cripps, E. (2010). Saving the polar bear, saving the world: Can the capabilities approach do justice to humans, animals and ecosystems? *Res Publica, 16*(1), 1–22.
De Geus, M. (2009). Utopian sustainability: Ecological utopianism. In L. Leonard & J. Barry (Eds.), *The transition to sustainable living and practice.* (pp. 77–100). Cheltenham: Emerald.
Deb, D. (2009). *Beyond developmentality: Constructing inclusive freedom and sustainability.* London: Earthscan.
Dobson, A. (2009). Freedom and dependency in an environmental age. *Social Philosophy and Policy, 26*(2), 151–172.
Dryzek, J., & Schlosberg, D. (1995). Disciplining Darwin: Biology in the history of political science. In Farr, D., & Leonard, S. (Eds.), *Political science in history: Research programs and political traditions* (pp. 123–145). Cambridge: Cambridge University Press.
Ehrenfeld, D. (1978). *The arrogance of humanism.* Oxford: Oxford University Press.
Fineman, M. (2008). The vulnerable subject: Anchoring equality in the human condition. *Yale Journal of Law and Feminism, 20*(1), 1–24.
Friedman, B. (2006). *The moral consequences of economic growth.* New York, NY: Knopf.
Fromm, E. (1976). *To have or to be?* New York, NY: Continuum.
Gabrielson, T. (2008). Green citizenship: A review and critique. *Citizenship Studies, 12*(4), 429–446.
Gabrielson, T., & Parady, K. (2010). Corporeal citizenship: Rethinking green citizenship through the body. *Environmental Politics, 19*(3), 374–391.
Goodin, R. (1985). *Protecting the vulnerable: A re-analysis of our social responsibilities.* Chicago, IL: University of Chicago Press.
Gupta, V. (2009). Framing the collapsonomics practice. Retrieved from http://vinay.howtolivewiki.com/blog/global/framing-the-collapsonomics-practice-1338. Accessed on February 5, 2011.
Havel, V. (1978). The power of the powerless. In J. Vladislav (Ed.), *Vaclav Havel: Living in truth* (pp. 36–123). London: Faber and Faber.
Havel, V. (1986). *Living in truth.* London: Faber and Faber.

Havel, V. (1991, October 29). A politician needs principles and good manners. *International Herald Tribune*.
Hay, P. (2002). *A companion to environmental thought*. Edinburgh: Edinburgh University Press.
Hine, D. & Kingsnorth, P. (Eds.). (2010). *Dark Mountain: Issue 1*. Bodmin: The Dark Mountain Project.
Homer-Dixon, T. (2006). *The upside of down: Catastrophe, creativity and the renewal of civilization*. Toronto: Alfred A. Knopf.
Humphrey, M. (2007). *Ecological politics and democratic theory: The challenge to the deliberative ideal*. London: Routledge.
Jackson, T. (2009). *Prosperity without growth: Economics for a finite planet*. London: Earthscan.
Jensen, D. (2002). *The culture of make-believe*. New York, NY: Context Books.
Jensen, D. (2010). *Resistance against empire*. Oakland, CA: PM Press.
Kasser, T. (2008). Pain and insecurity, love and money. *Psychological Inquiry*, *19*, 174–178.
Keat, R. (1994). Citizens, consumers and the environment: Reflections on *The Economy of the Earth*. *Environmental Values*, *3*, 4.
Knight, K. (Ed.). (1998). *The MacIntyre reader*. Cambridge: Polity.
MacGregor, S. (2006). *Beyond mothering earth: Ecological citizenship and the politics of care*. Vancouver: University of British Columbia Press.
MacIntyre, A. (1999). *Dependent rational animals: Why human beings need the virtues*. London: Duckworth.
Marcuse, H. (1964). *One-dimensional man: Studies in the ideology of advanced industrial society*. London: Routledge and Keegan Paul.
Mellor, M. (1997). *Feminism and ecology*. Cambridge: Polity Press.
Meyer, J. (2001). *Political nature: Environmentalism and the interpretation of western thought*. Boston, MA: MIT Press.
Midgley, M. (1995). *Beast and man: The roots of human nature*. London: Routledge. Rev. ed.
Murdoch, W., & Connell, J. (1971). All about ecology. In I. Horowitz (Ed.), *The troubled consciousness: American social issues* (pp. 56–63). *Pala Alto, CA: James E. Freel and Associates*.
Nussbaum, M. (2004). Beyond 'compassion and humanity': Justice for nonhuman animals. In C. Sunstein & M. Nussbaum (Eds.), *Animal rights: Current debates and new directions* (pp. 299–320). Oxford: Oxford University Press.
O'Neill, J. (1993). *Ecology, policy and politics: Human well-being and the natural world*. London: Routledge.
Offer, A. (2006). *The challenge of affluence: Self-control and well-being in the United States and Britain since 1950*. Oxford: Oxford University Press.
Orlov, D. (2011). *Reinventing collapse: The Soviet experience and American prospects – Revised and updated*. Gabriola Island, Canada: New Society Publishers.
Patočka, J. (2007). *L'Europe après l'Europe, translated from the Czech and German by Erika Abrams and Edited by Marc Crépon*. Lagrasse: Éditions Verdier.
Pontuso, J. (2004). *Havel: Civic responsibility in the postmodern age*. Lanham, MD: Rowman and Littlefield.
Prentice, H. (2001, June). Keynote Speech for community psychology/race and culture group annual conference. Unpublished manuscript.

Salleh, A. (1999). *Ecofeminism as politics: Nature, Marx and the postmodern*. London: Zed Press.

Salleh, A. (Ed.). (2009). *Eco-sufficiency and global justice: Women write political ecology*. London: Pluto Press.

Schlosberg, D. (2007). *Defining environmental justice*. Oxford: Oxford University Press.

Shusterman, R. (2008). *Body consciousness: A philosophy of mindfulness and somaesthetics*. Cambridge: Cambridge University Press.

Singer, P. (1990). *Animal liberation* (2nd ed.), London: Jonathon Cape.

Smith, M. (2006). Environmental risks and ethical responsibilities: Arendt, Beck, and the politics of acting into nature. *Environmental Ethics*, *28*(3), 227–246.

Smith, M. (2008). Suspended animation: Radical ecology, sovereign powers, and saving the (natural) world. *Journal for the Study of Radicalism*, *2*(1), 1–26.

Somma, M. (2009). Return to the villages. In L. Leonard & J. Barry (Eds.), *The transition to sustainable living and practice* (pp. 1–21). Bingley, UK: Emerald Press.

Sontag, S. (1978). *Illness as metaphor*. New York, NY: Farrar, Straus & Giroux.

Suzuki, D. (1997). *The wisdom of the elders*. New York, NY: Bantam Doubleday Dell Publishing Group.

THE TRANSITION TO GREEN LIFESTYLES BASED ON VOLUNTARY SIMPLICITY: THE DIFFICULT ROAD TOWARDS ENJOYABLE, GRACEFUL AND SUSTAINABLE LIFESTYLES☆

Marius de Geus

ABSTRACT

Purpose – *To explore the essential elements of a 'green art of living': an enjoyable, elegant, graceful and simultaneously low impact lifestyle.*

Design/methodology/approach – *The theoretical analysis is based on a critical reading and literature study of the texts of various prominent ecological utopian thinkers in the history of political philosophy.*

☆This chapter is a thoroughly revised version of my paper presented at the Symposium 'The Transition to Sustainable Communities: Justice, Resilience and Practices of Sustainable Living', 29 March 2010; School of Politics, International Studies and Philosophy, University of Belfast.

Environmental Philosophy: The Art of Life in a World of Limits
Advances in Sustainability and Environmental Justice, Volume 13, 31–49

ISSN: 2051-5030/doi:10.1108/S2051-5030(2013)0000013006

Findings – *In an ecologically sustainable society first priority must be given to re-examination of the proper 'ecological limits' of the current lifestyles and 'arts of living' in the Western world. The exact form or shape that people give to their lives is less important than their overall commitment to live within the ecological boundaries set by our earth.*

Originality/value – *A green, ecologically responsible society cannot do without a certain degree of moderation, self-restraint and simpler and less consumption-oriented lifestyles. However, in this chapter it is shown that such a society will not lead to impoverishment and austerity. An ecological society founded on voluntary simplicity will not be frugal or poor, but creates ample opportunities for its citizens to lead attractive, pleasurable, fulfilling and high-quality lives.*

Keywords: Utopia; ecological utopia; responsible lifestyles; ecological sustainability; green art of living; voluntary simplicity; moderation; self-restraint; excessive luxuries; living within limits; individual freedom; ecological footprint

INTRODUCTION

Modern mass 'consumerism' is increasingly considered to be one of the major causes of the global environmental crisis and our current climate problems.

Fundamentally, the various IPCC Assessment Reports (1990, 1995, 2001, 2007, 2013), *The Stern Review* (2006) and the *State of the World Report 2012* by the World Watch Institute have assessed that in the long run today's materialistic and energy-intensive lifestyles are simply untenable. Anyone who has studied these reports and publications such as James Hansen's *Storms of my Grandchildren* (2010) and Tim Jackson's *Prosperity without Growth* (2009) will come to the conclusion that the current levels of production and consumption in the Western world are simply unsustainable.

Hence, in our modern era the most urgent question has become: how can we all live comfortably and well without collectively exceeding climatic tipping points and ecological boundaries? (Wijkman & Rockström, 2012, Chaps. 3–6, pp. 52–85). In other words: how can an ecologically sustainable 'art of living' be developed in a world of limits?

From a political philosophical perspective this is without any doubt a crucial, but at the same time rather daunting question. Can we think of inspiring visions and strategies to convince the citizens of the Western world to reject their current energy intensive lifestyles and high level consumption patterns, and choose for credible transition paths into the direction of ecologically responsible forms of consumption?

The social, political, economical and cultural implications of this target are immense. This, however, is actually the immense task which must be fulfilled in the coming decades by our politicians and policymakers, the environmental movement, writers, journalists and political theorists.

As a matter of fact, the main challenge for political theory and politics will be to persuade the citizens of the West that it is inevitable to reject the dominant 'high energy' conceptions of 'the good life'. It seems, indeed, that there is no other way than to opt for essentially low carbon, joyful and creative lifestyles which are less based on materialistic pleasures. These innovative arts of living will imply more ecologically responsible behaviour, and will lead to a different level and alternative structure of consumption and economic production.

THE NEED FOR A RADICALLY NEW CONSUMER CULTURE AND LIFESTYLE CHANGES

In the professional political theoretical literature many hypotheses on the deeper causes of the modern ecological crisis can be found. Broadly speaking, in the eco-political writings of the last four decades there has been a focus on economic, social, scientific and technological factors that may explain the origins of today's environmental problems.

In critical economic analyses, the capitalist system is generally viewed as the main driving force behind the degradation of nature and environmental pollution. According to these analyses, in pre-capitalist societies human cooperation could still blossom and expansion and accumulation were not observed as the ultimate goals. When capitalism developed this radically changed. From then on the expansion of the economic system and accumulation of capital became predominant. Henceforth, trade and industry had to adapt to the capitalist ideology that dictated the imperative of 'grow or die'. Consequently, economic growth became vital and the natural world was reduced to a resource that could be exploited at will.

Prominent eco-philosophers such as Garret Hardin and William Ophuls, in contrast, have argued that most of today's environmental problems are caused by the mechanisms of collective behaviour and social dilemmas. In their analysis, free rider behaviour in a Commons brings ruin to all participants. Both citizens and trade and industry will try to profit from collective goods such as clean air and water, without actually paying for them. We are basically locked into a system of 'fouling our own nest', as far as citizens behave as independent, rational, free enterprisers. As long as citizens continue to behave as 'maximisers', the well known 'Tragedy of the Commons' will constantly reappear in problems of environmental pollution (Hardin, 1973, p. 139).

Techno-political thinkers like Hannah Arendt, Lynn White Jr. and Langdon Winner have put emphasis on the role of science and technology. According to these eco-philosophers, the basic roots of the ecological crisis lie in the progress of science, the development of new human attitudes towards technology and the longer term consequences of modern technologies. From the 17th and 18th centuries the natural sciences have contributed to the invention of new production techniques that have disrupted the delicate balance with nature and thereby contributed to environmental degradation. In combination with a nature-hostile attitude and a generally techno-optimistic friendly attitude, this development has caused irreversible global environmental risks, such as radioactivity and toxins and pollutants in the air, water and our food (Arendt, 1958; White, 1967, pp. 25–35; Winner, 1977).

Interestingly, Australian green philosopher Robyn Eckersley has pointed out more than twenty years ago that a growing number of eco-political writers have come to regard the global environmental problematic 'not only as a crisis of participation and survival, but as a crisis of culture in the broadest sense of the term, that is the total of the inherited ideas, beliefs, values, and knowledge, which constitute the shared basis of social action' (Eckersley, 1992, pp. 19–20). She has suggested that we approach environmental problems primarily as consequences of our Western culture and character, especially with respect to attitudes towards nature, lifestyles and ways of consumption (Eckersley, 1992, pp. 17–21). In this approach the overall structure of needs and lifestyles are questioned and the acquisitive values of today's consumerist society are fundamentally challenged.

At an earlier stage, deep green ecological philosophers like Arne Naess, Bill Devall and George Sessions have already underlined that most environmental dislocations are inherent to our Western culture of limitless expansion, economic growth, and our materialist and consumerist lifestyles (Devall & Sessions, 1985; Naess, 1989). However, if it is true that our deeply rooted norms, values, beliefs and aspirations are at the very root of today's environmental problematic, our basic ecological problems seem irremediable by single-issue economic reforms, scientific optimism and technological progress, or mere changes in rationality in order to prevent free rider behaviour leading to complex social dilemmas.

SOCIAL FACTORS INCREASING THE NECESSITY FOR RADICAL CHANGES IN LIFESTYLE AND CONSUMPTION

It is a fact of life that Western governments are insufficiently convinced or aware of the fact that the global environmental problematic is primarily a socio-cultural problem, despite the growing political-philosophical critique that what is essential is a drastic change in Western lifestyles, interfering deeply in our value systems, beliefs and consumption patterns. A lasting improvement of the state of the global environment and the quality of our natural habitat will require stringent measures for the producers of goods and services, *but* also for the consumers who – by internalising alternative consumption values and applying more simplified and responsible consumption patterns – will give direction to truly sustainable ways of living. How can it be argued that the consumer will increasingly have to play a key role in the abatement of environmental pollution and the realisation of long term ecological sustainability?

In *The State of the World 2012: Moving Toward Sustainable Prosperity*, published by the World Watch Institute, it is shown that there will be an ongoing growth of the world population to 9–11 billion by the year 2050. This will lead to a growing need for space and a trend for urbanisation that will in turn cause growing environmental pollution and the danger of depletion of scarce natural resources. In combination with a growing consumptive demand this will stimulate economic growth, an increase of energy use for production and transport, and an accumulation of the pressure on biodiversity and ecological stocks (World Watch Institute, 2012, pp. 129–137).

At the same time, Western cultural patterns with their emphasis on hedonism, their focus on property, luxury and convenience get more and more dispersed over the whole planet. The desires, aspirations, wants and needs of the affluent Western citizens are transmitted by mass media (TV, Radio, Internet) to all regions of the earth and further a demand for energy-intensive consumer goods (cars, electrical appliances), services and lifestyles.

In general, in the sixties, seventies, eighties and nineties of the twentieth century, there has been a considerable increase in family incomes in the Western world due to welfare growth and also the double incomes resulting from women starting to actively participate in the labour process. By this growth of income the consumptive spending per person has gone up considerably in those years. Despite the Western economic and financial crisis since 2008, it is expected that in particular energy intensive spending in the fields of housing, free time and holidays will rise strongly in the future. In the developed countries especially air traffic and high speed rail traffic will be on the increase, while in the lower income countries there will be a spectacular upgrade of car mobility (China, India, Brazil, etc.). Strikingly, a 'quadrupling' of world-wide person mobility until the year 2050 is to be expected (World Watch Institute, 2012, pp. 53–66).

Equally important are research findings indicating consumer behaviour to be increasingly guided by a wish and desire for comfort and convenience. This particularly applies to the domains of housing, recreation and nourishment (e.g. increased meat, fish and dairy product intake). A main complication here seems to be that most consumers expect environmental problems will be solved in thirty or forty years by technological means. It can be seriously doubted, though, whether the technical solutions needed will and can occur without radical changes in consumption styles, consumer behaviour and 'arts of living'.

Paradoxically, although individual consumer spending is increasing, human needs and desires have not decreased. As was already explained by Thomas Hobbes (1588–1679) in his *Leviathan*:

> Felicity is a continual progress of the desire, from one object to another; the attaining of the former being still but the way to the latter. The cause whereof is, that the object of man's desire, is not to enjoy once only, and for one instant of time; but to assure for ever the way of his desire. And therefore the voluntary actions, and inclinations of all men, tend, not only to the procuring, but also to the assuring of a contented life. (1974, p. 161)

A fundamental dilemma of modern liberal democratic society is that the satisfaction of one consumptive need tends to lead to the creation of other desires and wants among consumers: because of the apparently insatiable needs of modern humans a situation of abundance is beyond reach, and over and over again this produces new forms of 'scarcity'.

Taking these specific social factors into consideration, it will be clear that nowadays a profound attention to the contribution of individual consumers to environmental pollution and the degradation of nature is essential, more than ever before. In the last two decades this has provoked an intriguing theoretical debate on whether or not in our era the time has come to place the environment above individual pleasure and enjoyment caused by excessive levels of hedonistic consumption.

LIBERAL DEMOCRACY AND THE REDUCTION OF CONSUMERISM

In this context, it is particularly important to understand why up to this day Western liberal democracies have been reluctant and incapable to introduce policies that would restrict the citizens' freedom of choice. It is fair to say that despite the fact that Western consumption patterns are often interpreted as the most significant cause of environmental pollution and nature degradation governments in the Western world have not given priority to a reduction of consumerism, more general decreases in individual levels of material affluence, or changes in lifestyles.

Obviously, Western governments that are themselves addicted to continuous economic growth and expansion will not be inclined to induce their citizens to decrease their spending and consumption levels. In many ways economic growth has become a necessary condition for the flourishing of the modern welfare state. A constant growth has become necessary to facilitate an improvement of material prosperity of the citizens, to keep social expenditures affordable, and to solve distribution conflicts between the more and less affluent sections of society.

In order to keep the wheels of the economy going an adjustment of consumerist patterns of Western citizens is often seen as undesirable and even detrimental to the stability of the economy. Indeed, it is the relentless rise in consumer wants and demands that stimulates the economy and that makes our welfare state prosper, albeit in a highly restricted material sense.

Moreover, in Western liberal democracies status and identity are associated with investment in property and the possession of goods and products. In order to distinguish oneself from others, citizens are longing for possessions and the display of conspicuous forms of consumerism. Until recently, a lifestyle of abundance was only within reach of a privileged elite, but today ever larger groups in society owe their sense of identity and self-respect to material affluence, consumerism and status goods.

After the Second World War, personal self-restraint and moderation lost their place as core values in the West as times of wealth, luxury and plenty took hold. Hegemonic ideology transmitters like MTV and TMF teach our children that spending more can be a 'liberating experience' and the luxurious lifestyles of pop stars and other idols have become the central goal to strive for in life. Their promise is that consumerism will provide us with meaning, respect, status, power, fulfilment, distraction and enjoyment: spending money, living a luxurious life and enjoying spectacular vacations in far away holiday resorts have gradually become the norm for many affluent citizens, both young and old.

According to prominent American sociologist Juliet Schor in *The Overworked American* (1993), *The Overspent American* (1998), *Born to Buy: The Commercialized Child and the New Consumer Culture* (2004), and *Plenitude: The New Economics of True Wealth* (2010), many middle class and upper-middle class people are victims of 'the insidious cycle of work-and-spend'. For them there is a strong incentive to work longer and earn more in order to consume more resources and enlarge their possessions. In the Western world the value of work has gradually risen and the value of free time has gone down (Schor, 1993, pp. 107–112). People are increasingly prepared to work more since work has become the vital instrument that will make it possible to consume their way to happiness.

To be sure, many citizens are trapped in the illusion of having to keep up with the often mentioned 'Joneses': their family, neighbours and friends. This may explain why over the last five decades – despite increased productivity – the average work week for many American citizens has gone up considerably. Americans often hold more than one job and are working more overtime in order to meet consumer expenses and pay off their debts (Schor, 1993, pp. 28–31, see also Schor, 1998, 2004).

A major concern is that on a societal level leisure and free time have decreased significantly, while the happiness of the population as a whole has gone down: people get worn out, stressed and even depressed by the high work load and the fact that eventually the enticing promises of full-fledged consumerism often turn out to be meaningless. Western liberal democracies have always stressed that the freedom to consume keeps the capitalist economic system going and constitutes an inalienable right of the individual citizen: the freedom to consume can be seen as the basic expression of the Lockean creed of valuing 'life, liberty and the pursuit of happiness'. In this hegemonic liberal democratic ideology the role of the government is certainly not to restrict consumer behaviour or to re-evaluate personal lifestyle choices or 'arts of living' in the light of the requirements of an ecologically sustainable development.

GREEN LIFESTYLES AND VOLUNTARY SIMPLICITY

I now turn to a number of intriguing issues concerning a green art of living based on the idea of voluntary simplicity, and evaluate the various aspects of this sustainable alternative. Elsewhere, particularly in my *The End of Over-Consumption* (2003), I have argued that in modern Western democracies personal and social status and identity are basically looked for in property. In order to distinguish oneself from others, citizens are longing for possessions and displaying conspicuous forms of consumerism. Until recently, a lifestyle of abundance was only within reach of a small elite, but as mentioned above nowadays ever larger groups in society owe their sense of identity and self-respect to material affluence and consumerism. *Grosso modo*, in the Western world personal self-restraint and moderation are no longer core values in times of wealth, luxury and plenty.

In a world dominated by the idea of systematically consuming our way to identity and happiness, and by the constant status anxiety which exists on a societal level of not being able to keep up in the rat race with one's fellow citizens, it may seem surprising to accentuate the positive sides of a voluntary simplicity, and an art of living based on temperance and restraint. However, in the personal lives of myself and many fellow citizens which I have met over the years during dozens of speeches and lectures, it turns out that moderation of consumption and a simplification of our lifestyles can actually be highly beneficial. This is evident in terms of living more consciously and directly, living closer to nature, and especially having more time and energy to live responsibly and enjoy life to the fullest.

What, then, are the most significant characteristics of living an ecologically sound life which is essentially based on the concept of voluntary simplicity? The starting point for an alternative ecologically responsible and simplified lifestyle and green art of living is the avoidance of clutter and excessive luxuries. As noted above, many modern citizens in the Western world are obsessed with continuously buying new consumer goods.

As Tim Jackson has explained in his ground breaking *Prosperity without Growth; Economics for a Finite Planet* (2009), this constant focus on 'novelty' has been one of the most significant undermining forces of ecological sustainability: 'The production and consumption of novelty drives the growth economy. Novelty both reinforces and is reinforced by the social logic of consumerism' (Jackson, 2009, p. 202).

Yet today's unstoppable desire for new and innovative products is having many far reaching consequences. In our economic system new and exciting consumer goods continually emerge and make existing products like cars and computers quickly look outdated and unattractive. Producers are continually trying to stimulate consumer demand and many of us have indeed become addicted to purchasing new technologies and 'sexy' products like sleek smart phones, tablet computers and fancy digital cameras. In consequence, the older consumption goods that people own – and which in many cases are still fully functional – tend to become obsolete prematurely.

The 'throw-away society' or – more precisely – the 'clutter-society' are indications of one and the same basic problem: countless items and goods which are still highly functional are often remorselessly and at high material and energy costs being replaced by novel products. In fact, it is telling that currently most Western houses are full of clutter, often littered in living and bed rooms, and stored in cellars, attics, garage boxes and sheds.

An art of living based on voluntary simplicity, then, is in my view best described as a systematic avoidance of exterior clutter and by making use of products in environmentally responsible ways, as long as they actually are functional. This constitutes a relatively easy and transparent strategy to achieve a more environmentally friendly and sustainable lifestyle. When buying new products consumers must consistently take into consideration whether the goods they already own are still satisfactorily fulfilling their purpose. If this is the case, it is essentially 'anti-ecological' to purchase a new product or good only out of curiosity or a relentless desire for novelty. This is a straightforward and appropriate criterion which is relatively easy to apply in daily practice and which will lead to more frugal and ecologically responsible forms of consumption.

The critical point to note here is that in general terms the concepts of temperance, moderation and frugality with their often negative connotations seem incompatible with our current culture of mass consumerism. However, for instance the term frugality has often been misunderstood as has been contended by *Financial Times* columnist Harry Eyres:

> (...) the words for frugal in English and Roman languages derive from the Latin frugi, an indeclinable adjective formed from the dative of frux (fruit), and often combined with bonae – so 'to or for the good fruit'. Being 'for the good fruit' means being honest and temperate, dedicated to long term flourishing: as vital for human beings as for the earth itself. (Harry Eyres, quoted in Jackson, 2009, p. 192)

To be frugal implies getting satisfaction and joy from using consumer goods in mindful ways and for sensible purposes. Surely, if people are systematically and constantly in need for the consumption of novelty products this obviously cannot be called an ecological way of living. In sum, an art of living stemming from voluntary simplicity means to live a life of purpose, which is considerate to the natural environment, clutter free and also materially light.

Given the centrality of reduced and also 'different forms of' consumption (also see below) it must be emphasised that voluntary simplicity does not imply the rejection of every form of excess and luxury, or to choose for an unadorned, sparse and even poor life. In this respect, I would suggest the solution for the modern environmental crisis must be found in maintaining a 'comfortable prosperity', in combination with moderate forms of luxury. In this sense, enjoying certain luxuries is definitely not some ecological sin, but is acceptable as long as the ecological consequences are kept within reasonable limits.

No doubt the concept of 'moderate luxury' excludes a full-fledged consumerism that takes the form of opulent excess, that is overly hedonistic, narcissistic and materialistic. Clearly, it does not mean to deny individuals having pleasant experiences, but it aims to prevent excessive luxury consumerism, extravagant lifestyles and opulent spending on luxury goods. For instance, it would exclude very thirsty four-wheel drive cars like Hummers, energy inefficient jet skis and snow scooters when used only for recreational purposes, or frequent intercontinental flying to exotic holiday areas.

In this line of reasoning the central goal is to achieve a well-studied, comfortable lifestyle, by realistically simplifying our consumption patterns so that nature and the environment can be effectively protected and the planets' limits are respected. Its core idea is that as long as unnecessary waste in production and consumption are prevented, 'reasonable' forms of luxury in combination with pleasant and enjoyable living conditions are the best overall strategy. Of course, these 'reasonable' luxuries should best be selected on the basis of their potential for saving work and energy, and must be characterised by optimal efficiency, for example by using plug-in hybrid cars or fully electrical cars powered by renewable, 'green' energy, or extremely energy efficient notebooks, desktop computers, LED television sets, etc.

SLOWING DOWN, ENJOYING FREE TIME AND WORKING LESS

In the past, prominent ecological philosophers such as Henry Thoreau, Peter Kropotkin, William Morris, Ernest Callenbach, Ivan Illich and

Murray Bookchin have stressed the need for a fundamental revaluation of a slower pace of life, leisure time and rest (Also De Geus, 1999, p. 212). In an ecologically viable society based on voluntary simplicity, haste will have to make way for a slower and calmer work tempo, restlessness for peacefulness, and lack of time for enough leisure time. In this view, working long and hard must be replaced by recreation, the chance to relax and to actually have a lot of time for oneself.

For instance William Morris (1834–1896) has pointedly given the subtitle '*An Epoch of Rest*' to his most inspiring utopian novel *News from Nowhere* (1891). He empathically criticises the degradation of nature and environmental pollution and condemns society's exaggeratedly fast pace, which causes people to always be stressed trying to keep up. In his perspective, the quality of life is essentially determined by the chance to calm down, to relax and apply oneself to self-development and community life. Morris is one of the best examples of critical social theorists who have referred extensively to the value of leisure time for being able to develop one's artistic, creative and intellectual talents and activities.

In line with this American philosopher Jerome M. Segal has underlined in his engaging *Graceful Simplicity* (1999) that an ecologically sound and good life is incompatible with a hurried lifestyle. He argues that overall the use we make of time is a quintessential problem: 'The good life requires true leisure: not just time that we are not at the job, but time that is free from pressing demands. We seek freedom from an excessive amount of both paid and personal work' (Segal, 2003, p. 24). Again, it is important to note that a simpler and less consumerist lifestyle and art of living do not necessarily imply a boring or quiet existence:

> It can be filled with challenge and excitement. But it does give recognition to the importance of having sufficient time to go slowly, and to do things right – whether we are eating a meal, talking with a friend, telling a story to a child, or walking the dog. (Segal, 2003, p. 24)

Overall, citizens who opt for voluntary simplicity and green lifestyles are people who work less, who focus less on paid work and consume less. Notably, they have enough time to choose for alternative ways of consumption which indeed in some cases may take more time, like when using your bike or public transport instead of taking the car, repairing a consumer item by yourself, or preparing a freshly home-made meal instead of buying a frozen, instant dinner.

GRACEFULNESS AND ELEGANCE VERSUS AUSTERITY

A vital issue to be confronted by political philosophers and policy makers is whether an art of living stemming from voluntary simplicity, will be based on either the concept of 'austerity' or 'gracefulness'. In the early days of the environmental movement the concept of austerity has often been put forward as a green solution for over-consumption and excessive spending. At best, austerity may mitigate some of the most immediate symptoms of our mass consumerist society, but apparently in the long-term citizens will be feeling uncomfortable with extreme forms of abstinence from consumerism and abundance. Undoubtedly, there is no need to choose for a most radical interpretation of austerity: to refuse every form of excess and luxury, and to choose a life that is entirely unadorned, sparse, restricted and even poor.

Ultimately, it will be counterproductive to look for the solution to our current environmental and climate problems in forms of often religiously inspired 'personal ascetism' and extreme forms of self-restraint. Indeed, in some cases it may make sense to argue for a limitation of particular material needs and luxury wants such as using four-wheel drive SUVs in flat cities or countries, but from a strategic point of view it seems unwise to consistently choose for a minimal standard of living and a radically low level of consumption.

Generally, in modern green political thinking, an ecologically sound society does not by definition imply a barren and meagre society, nor will it inevitably be leading to an ascetic lifestyle. On the contrary, a large majority of green thinkers argue for maintaining a 'comfortable prosperity', while at the same time proposing a whole series of far-reaching policy measures to improve the eco-efficiency of goods and services, to reduce emissions and the use of scarce materials, and to protect the natural environment in the most effective ways.

In contrast with the traditional view of austerity, I argue for a different interpretation of austerity, namely an 'artistic one'. In this perspective, the term austerity does not have the connotation of scantiness, scarcity and poverty, but one of elegance and gracefulness. Accordingly, living a life of elegance and gracefulness implies living with style, based on the principles of 'simple beauty', and of sound and aesthetically pleasing designs and structures (of goods, clothing, housing, city buildings and so on).

In this line of thinking there is room for 'a sense of moderation without impoverishment'. On this reading, a society which is founded on an art of living emanating from voluntary simplicity is not frugal or poor, but makes possible an attractive life that is well provided and cared for, while not extravagant or wasteful. This alternative art of living will be based on an economy of enough, in which elegance and aesthetics are of primary importance.

In this context, Jerome Segal has convincingly referred to the point that nowadays only few people in the Western world are actually living the good life: 'Instead we live in a frantic pace. We live lives filled with anxiety, filled with hassles and headaches and hard edges. What is most absent from our lives is gracefulness' (Segal, 2003, p. 249). In his analysis, gracefulness must be understood in terms of 'beauty, an aesthetic of time, inner peacefulness, appreciativeness, and generosity of spirit':

> To live gracefully is to have mastered an art of living – and as with most art, true mastery requires inner development. Yet while achieving gracefulness is accomplished one life at a time, no social environment is neutral with respect to gracefulness. The social and economic world we inhabit either facilitates graceful living or frustrates it. (Segal, 2003, p. 249)

In the future an eminent political philosophical issue will then be how to create plenty of room for and stimulate an array of elegant, graceful, simplified and sustainable lifestyles which will cause no irreversible dangers to our much cherished planet.

SYSTEMATICALLY INVESTING IN HIGH TECH GREEN PRODUCTS AND CONSUMING 'DIFFERENTLY'

In general terms, social critics of our modern consumerist society have underlined the need for radical lifestyle changes and a drastic reduction of the levels of consumption. However, the core problem here is not so much to stimulate people to consume drastically less, but to continuously invite and require them *to consume in more environmentally friendly and sound ways.* It must be accentuated that consumerism as such is not by definition inherently egoistic, materialistic or just another deplorable indication of greed and the search for status and prestige. We must accept that it is virtually impossible to live a good life without the production and consumption of certain goods and services, such as food, shelter and a number of basic tools. This significant point has been earlier made by John Barry in his book *Rethinking Green Politics*:

> Green political economy does not seek to reject individual material consumption as unworthy, or indeed as necessarily un-ecological. (...) Although frugality and simplicity of lifestyle may have their own virtues, a *moderately* materialistic lifestyle need not be without its own inner rewards, and ecological virtues. But these virtues will not be realized unless consumption is integrated as an integral aspect of ecological stewardship. Material consumption becomes an ecological vice where it orientates itself around nothing other than consumption itself. The resolution of the tension between frugality and excessive consumption lies in an admittedly imprecise, middle position of optimality or sufficiency. (Barry, 1999, p. 176)

I do concur with Barry that the key issue revolves around finding the right combination or mix of technological innovation (producing high tech environmentally sound 'green' products such as extremely efficient heating, refrigerators, washing machines, etc.), and *different* types of consuming goods and services by us buyers, not necessarily accompanied by much lower standards of living. Increasingly, it will involve responsible forms of consumption by orientating on consumption goods which are optimally eco-efficient, and also re-usable, easy to take apart and repair, easy to recycle and renovate as well, as to upgrade by for instance replacing specific parts or the installation of new software.

Thus, nowadays decisive progress can be made by the introduction of high technological domestic heat pumps, superefficient solar cells on our homes, small high-quality domestic roof-mounted wind turbines, and introducing various structural changes (e.g. prevention, closing recycling loops, and particularly 'upcycling': converting waste materials into products of higher environmental value. In this way, we will be able to substantially reduce harm to the environment in the future, while still being able to maintain prosperity at a high level. As a matter of fact, in the recent *Global Environment Outlook 2012* by the United Nations (GEO 5) there is argued for a 90% (Factor 10) reduction in resource consumption for ecological reasons, if adequate resources are to be released for the needs of the developing countries' (United Nations Environment Program [UNEP], 2012).

This revolutionary goal would imply a systematic improvement in the efficiency of resource consumption and production processes. In fact, the 90% reduction target will greatly affect our heating, cooling and transport systems in order to conserve the natural environment while the standard of living remains the same or may increase in specific areas. Obviously, this project would also include the newest 'zero-energy' or even 'energy producing' homes which embody the highest possible degree of efficiency, economy and comfort. This general direction of solution seeking by opting for voluntary simplicity, including consuming and producing in *fundamentally different ways*, offers feasible solutions and in my analysis deserves further consideration and scientific research.

Last but certainly not least, there is the determining question if green lifestyles based on voluntary simplicity may constitute an imminent danger to individual citizens' rights and the great variety of lifestyles in modern Western countries. In general, most people would agree that a great diversity of lifestyles is a fact of modern life and a 'good' to be cherished in society. Surely, I totally agree with that specific point of view. However, what is also needed, is a fundamental re-examination of the proper 'ecological limits' of the variety of lifestyles and 'arts of living' which are currently enjoyed.

Eventually, the precise shape that people give to their lives and their specific ideas of the 'art of living a good and green life' are less important than their commitment – regardless of the specific lifestyle they choose – to be living within the borders set by the carrying capacity of the Earth. By whatever means, individual citizens should now aim to keep their so-called 'ecological footprint' within sustainable limits.

In this context it is right to note that defining ecological limits does not only have a negative side. Often, limits have positive effects and may help to increase creativity and stimulate energies, and create new challenges for individuals, groups, organisations and societies as a whole, as has been convincingly argued by Wolfgang Sachs by the turn of the last century:

> Limits have a double nature, being both restraining and facilitating; they act as constraints only with respect to one order of things, but open up possibilities with respect to another order of things. (Sachs, 2000, pp. 179–180)

This suggests the real question concerns the extent to which, individually and collectively, we come to acknowledge the impact of our behaviour on the Earth and start to adjust our everyday material consumption in the light of global ecological constraints and limits. Naturally, this would require us to understand the approximate amount of environmental space or the 'ecological footprint' that we use, and to be prepared to reduce our overall claims on land, energy and other natural resources.

This is logically implied by the call from Matthis Wackernagel and William Rees for a so-called 'fair Earthshare' of 1.8 hectares for every inhabitant of the planet. They have defined this as 'the amount of land each person would get if all the ecologically productive land on Earth were divided among the present world population' (Wackernagel & Rees, 1995, p. 53). In any case, provided that people do not irresponsibly exceed their fair share of the Earth, ecologically responsible lifestyles based on voluntary simplicity definitely need not be one-dimensional or bleak, but offer their own new possibilities, rewards and excitements.

CONCLUDING REFLECTIONS

Finally, I turned to the core characteristics of a green art of living based on the idea of voluntary simplicity. In my analysis, the starting point for an alternative ecologically responsible and simplified lifestyle and green art of living, is keeping away from clutter and excessive luxuries. In addition, in an ecologically viable society based on voluntary simplicity, haste will have to make way for a slower and calmer work tempo, restlessness for peacefulness, and lack of time for enough leisure time.

In this line of thinking there must be room for 'a sense of moderation without impoverishment'. On this reading, a future green society which is founded on voluntary simplicity is by no means frugal or poor, but makes possible an attractive life that is well provided and cared for, while not extravagant or wasteful.

It was then discussed whether an ecological art of living stemming from voluntary simplicity, must be based on either the concept of 'austerity' or 'gracefulness'. A general perspective was prioritised in which austerity does not have the association with scantiness, scarcity and poverty, but instead with elegance and gracefulness. Thus, living a life of elegance and gracefulness implies living with style, based on the idea of 'simple beauty', and of aesthetically pleasing designs. Eventually, the major issue is to find the optimal combination of high tech green innovation, and *different* types of consumption by buyers of goods and services.

What is most needed, is a fundamental re-examination of the proper 'ecological limits' of the variety of lifestyles and 'arts of living' which are currently enjoyed. Principally, the precise shape that people give to their lives and their specific ideas of the 'art of living a good life' are less significant than their overall commitment to live within the ecological limits and boundaries set by our earth.

In sum, it will not be a particularly easy task to convince Western citizens to reject their structurally high carbon lifestyles and consumerist attitudes and patterns, and to choose for transition paths into the direction of an ecologically responsible art of living. It has become clear that living an ecologically sound life which emanates from the concept of voluntary simplicity will indeed be demanding and challenging. Nevertheless, a voluntarily chosen simple lifestyle will allow us to become masters of pleasures instead of becoming its slaves. It will make it possible for people to escape from the manifold temptations of our abundant consumerist society, and find fulfilment in a lifestyle which is more modest.

An enjoyable and attractive green new art of living will demand a permanent individual and social learning process that seeks to integrate liberty and diversity of lifestyles with specific social-ecological and moral responsibilities towards our fellow humans and descendants. Ultimately, this can lead to an ecologically sustainable society which is enjoyable, elegant, graceful and intrinsically rewarding, without exceeding global limits and our fair shares of the earth's resources.

REFERENCES

Arendt, H. (1958). *The human condition.* Chicago, IL: The University of Chicago Press.

Barry, J. (1999). *Rethinking green politics.* London: Sage Publications.

De Geus, M. (1999). *Ecological Utopias: Envisioning the sustainable society.* Utrecht, Netherlands: International Books.

De Geus, M. (2003). *The end of over-consumption: Towards a lifestyle of moderation and self-restraint.* Utrecht, Netherlands: International Books.

Devall, B. & Sessions, G. (1985). *Deep ecology: Living as if nature mattered.* Layton, UT: Gibbs M. Smith.

Eckersley, R. (1992). *Environmentalism and political theory: Toward an ecocentric approach.* London: UCL Press.

Hansen, J. (2010). *Storms of my grandchildren.* London: Bloomsbury.

Hardin, G. (1973). The tragedy of the commons. In H. Daly (Ed.), *Toward a steady state economy.* San Francisco, CA: Freeman.

Hobbes, T. (1974). *Leviathan.* Harmondsworth: Penguin Books.

IPCC Assessment Reports. (1990, 1995, 2001, 2007, 2013). Geneva: IPCC.

Jackson, T. (2009). *Prosperity without growth.* London: Earthscan Publications.

Morris, W. (1983). *News from nowhere, or an epoch of rest.* London: Routledge.

Naess, A. (1989). *Ecology, community and lifestyle.* Cambridge: Cambridge University Press.

Sachs, W. (2000). *Planet dialectics: Explorations in environment and development.* London: ZED Books.

Schor, J. (1993). *The overworked American.* New York, NY: Basic Books.

Schor, J. (1998). *The overspent American.* New York, NY: Basic Books.

Schor, J. (2004). *Born to buy: The commercialized child and the new consumer culture.* New York, NY: Scribner.

Schor, J. (2010). *Plenitude: The new economics of true wealth.* New York, NY: The Penguin Press.

Segal, J. (2003). *Graceful simplicity: The philosophy and politics of the alternative American dream*. Berkeley, CA: University of California Press.

Stern, N. (2006). *The Stern review*, London: HM Treasury.

UNEP. (2012). *Global environment outlook report [e Book]*. New York, NY: UNEP Publishers.

Wackernagel, M., & Rees, W. (1995). *Our ecological footprint: Reducing human impact on the earth*. Gabriola Island, Canada: New Publishers.

White, L. Jr. (1967). The historical roots of our ecological crisis. In I. Barbour (Ed.), *Western man and environmental ethics*. Reading, UK: Addison Wesley.

Wijkman, A., & Rockström, J. (2012). *Bankrupting nature: Denying our planetary boundaries*. London: Routledge.

Winner, L. (1977). *Autonomous technology: Technics-out-of-control as a theme in political thought*, Cambridge, MA: MIT Press.

World Watch Institute. (2012). *State of the world report*. Washington, DC: Island Press.

THE POLITICS OF GLOBAL WARMING (1): CLIMATE SCIENCE AND SCEPTICISM

Graham Parkes

ABSTRACT

Purpose – *The purpose of this chapter is twofold. First, to give a concise account of the current global climate situation, its previous history according to the palaeoclimate record, and climate scientists' predictions of the consequences of various scenarios of global climate change. Then to explain why so many people continue to be oblivious to the enormous risks of continuing with business as usual.*

Methodology/approach – *The approach is through a comprehensive study of the relevant evidence and the scientific and scholarly literature, interwoven with philosophical reflections on their significance.*

Findings – *The findings are as follows: the evidence for the anthropogenic nature of global warming is overwhelming, and the prognoses for continued burning of fossil fuels (sea level rise, extreme weather, etc.) are dire. The denial stems in large part from the undue influence of climate scepticism movements, lavishly funded by the fossil fuel industries, combined with a variety of psycho-social and economic factors.*

Environmental Philosophy: The Art of Life in a World of Limits
Advances in Sustainability and Environmental Justice, Volume 13, 51–80

ISSN: 2051-5030/doi:10.1108/S2051-5030(2013)0000013007

Social implications – *The implications are several. Given the complex nature of global warming, scientists need to do a better job of communicating their findings to the general public, and scholars and academics need to find ways to expose the machinations of the fossil fuel industries. And given the global impact of climate change, citizens of the developed nations need to see that a radical change in their behaviour is demanded not only by considerations of social justice but also even by their own self-interest.*

Originality value – *The value of this philosophical approach is that it affords a more comprehensive view of the situation around global warming than we get from the more specialised disciplines.*

Keywords: Climate change; climate scepticism; climate science; economics; global warming; politics

INTRODUCTION

Globalisation appears to have maddened the world, driven its citizens insane. By burning fossil fuels to provide energy for our high-comfort lifestyles, enjoying unprecedented mobility thanks to automobiles and planes, warmed in winter and cooled in summer, we in the developed world, and many in developing countries too, are driving the planet towards climate tipping points that could plunge civilisation as we know it into catastrophe. As millions in the affected countries lose their homes and livelihoods, there will be domestic chaos and mass migrations. Fortunately, the United States, which has contributed most to the warming and is responsible for blocking meaningful action to mitigate it, has lots of space for climate refugees.[1]

The earth has been under brutal assault for quite some time, suffering insult after insult at the hands of the human race – to the point now where our impact on the planet and its climate threatens to destroy a good portion of the natural resources on which we depend for our very survival. From the perspective of the world's classical philosophies, the wisdom of the ancients in most cultures, the current behaviour of most citizens in the so-called developed world looks like collective insanity, self-destructive madness, massive self-deception resulting in an ignorance so wilful as to be culpable. But those who are inflicting the damage seem incapable of restraining themselves, and the politicians appear not to care about anything beyond their next chance for re-election.

But perhaps they know not what they do. If people knew what is happening on the scale of the planet's climate, as a result of human activity, would they not want to avoid perpetrating a catastrophe that threatens the livelihoods of millions, as well as future generations even of the rich and fortunate? And even if, as many climate scientists now think, we have already altered the atmosphere so much that destructive consequences are unavoidable even if we change our ways immediately, it still makes sense for us to mitigate the ill effects as much as possible. Hope lies in people's coming to face the facts of the situation, to understand the meaning of what we are doing and how we are behaving – and from there to take action to change our collective behaviour.

As the evidence for anthropogenic climate change has accumulated, and scientific theories of how this process works have become increasingly well corroborated, much of the general public – especially in the United States – doubts in its wisdom that human activity is the major cause of the problem.[2] They say that ignorance is bliss, but the strange thing here is that the unrestrained consumption of energy and goods by the developed societies is by no means enhancing our happiness or health. A number of socio-psychological studies have shown that accumulating more money and possessions than we need (and especially in a context of impoverished relations to the earth and the places we inhabit) actually depresses and impairs human flourishing.[3] Higher levels of consumerism appear to correlate better with a rise in consumption of pharmaceuticals than with levels of human fulfilment.

The inability of short-sighted politicians to undertake longer range planning for the sake of the nation contrasts starkly with the attitude of the military, which has an enormous stake in being prepared. While the US government refuses to take action on climate change, the attitude of the Pentagon – hardly a hotbed of tree-hugging dreamers – is aptly realistic. The *Quadrennial Defense Review Report* from 2010 acknowledges the reality of climate change and the enormous challenges it poses. For one thing, the Department of Defense is disconcerted by the amount of military infrastructure at risk because of changing weather patterns: 'In 2008, the National Intelligence Council judged that more than 30 U.S. military installations were already facing elevated levels of risk from rising sea levels' (U.S. Department of Defense, 2010, p. 85). The Pentagon also perceives the more global threats it faces in a warming world, and is preparing for the protection of its military installations abroad.

> Assessments conducted by the intelligence community indicate that climate change could have significant geopolitical impacts around the world, contributing to poverty,

> environmental degradation, and the further weakening of fragile governments. Climate change will contribute to food and water scarcity, will increase the spread of disease, and may spur or exacerbate mass migration. (p. 85)

No need for that qualified 'may' in the last clause: climate refugees, or environmental migrants, are already on the move in the millions, and their numbers are growing. These people are fleeing homes inundated by flooding or destroyed by hurricanes, fields made barren by encroaching salt, crops withering from lack of water. The suffering from displacement from their homes is compounded by violent conflicts over resources. In the words of the *Pentagon Report*:

> While climate change alone does not cause conflict, it may act as an accelerant of instability or conflict, placing a burden to respond on civilian institutions and militaries around the world. In addition, extreme weather events may lead to increased demands for defense support to civil authorities for humanitarian assistance or disaster response both within the United States and overseas. (p. 85)

You bet: a big job ahead. There are always conflicts over resources, and when climate change makes these scarcer, this intensifies the antagonisms. Resource wars are becoming climate wars: convincing scenarios have been projected.[4]

While the politicians in Washington deny and prevaricate, the military at least 'talks the talk':

> The Department is increasing its use of renewable energy supplies and reducing energy demand to improve operational effectiveness, reduce greenhouse gas emissions in support of U.S. climate change initiatives, and protect the Department from energy price fluctuations. The Military Departments have invested in noncarbon power sources such as solar, wind, geothermal, and biomass energy at domestic installations and in vehicles powered by alternative fuels, including hybrid power, electricity, hydrogen, and compressed national gas. (p. 87)

Good work – as long as they are actively reducing their dependence on carbon energy sources. If only U.S. climate change initiatives were more aggressive, the Department of Defense would do even better.

Up to now, those who are hardest hit by the consequences of our abuse of the earth are not the worst abusers among the richer nations but, in a truly tragic irony, those poorer countries who have contributed least to the on-going destruction. And since the attempts of the developing nations to negotiate just and fair political and economic resolutions are constantly being rebuffed (as in the climate summit fiasco in Copenhagen in 2009), we can expect to see an increase in eco-terrorist attacks in the richer nations designed to draw attention to our complicity in worsening the plight of the poorer. Even if we are spared such terrors, the number of climate refugees will swell to many millions: we shall have been responsible for their parlous condition and will be obliged to take care of them. A climate scientist in Bangladesh has worked out a formula: if you tell him the size of your 'developed-world' carbon footprint, he can tell you how many Bangladeshis will be coming to live with you when their homes are submerged by the sea.[5] It is thus very much in our own self-interest to desist from disrupting the planet's climate – quite apart from the fact that to refrain is likely to make us a good deal happier.

What often seems to conduce to happier lives is the wisdom of indigenous peoples who understand the part of the earth they live on and know how to treat it appropriately. If you know your local climate, you can step outside in the morning and tell from various signs how the day's weather is likely to be. But this kind of understanding is of little direct help in the context of climate change, since the problem is global prior to its local manifestations. We cannot say what is happening to the global climate over the longer term without the aid of natural science, which stretches our understanding based on human lifetimes, using palaeoclimatology to cover ages of the earth that are measured in many millions of years.

CLIMATE SCIENCE

In 2011 the human race broke two world records. First, we discharged an unprecedented amount of carbon dioxide into the atmosphere, mainly by burning fossil fuels: almost *35 billion* tons (Global Carbon Project, 2012). This maximum was matched, not coincidentally, by a further rise in carbon dioxide levels in the atmosphere – to a record level of over 390 parts per million (ppm) (Peters et al., 2011).[6] And now the latest measurements of atmospheric concentrations of CO_2 show the rise continuing to accelerate: the increase during 2012 of 2.67 ppm is the second highest jump ever recorded, the record being 2.93 ppm in 1998 (Vidal, 2013). These trends suggest that we are generating an increase in global average temperatures that is already destroying the livelihoods of many people, and will soon jeopardise the lives of hundreds of millions more.

The connection between carbon dioxide and warming comes as no surprise, since scientists have understood for some time now how the composition of the atmosphere affects temperatures on the earth's surface. In 1824, the French mathematician and physicist Joseph Fourier likened the atmosphere to a gigantic bell-jar that prevents some of the sun's heat that reflects off the earth from escaping out into space (Fourier, 1824; discussed in Christianson, 1999). Some 30 years later, an Irishman by the name of John Tyndall discovered the processes that are behind this phenomenon, often referred to as 'the natural greenhouse effect'. Tyndall was one of the most versatile scientists of his time, and his interest in the ways various gases absorb and transmit radiation led him to build in 1859 the world's first ratio spectrophotometer.[7] With this device he was able to show that, while the oxygen and nitrogen that make up most of the surrounding air have almost no effect on radiant heat, the water vapour, carbon dioxide and ozone in the atmosphere absorb heat radiation. Heat-trapping gases are distinguished by their selectively absorptive properties, being transparent to the visible, short-wavelength, heat-imparting light of the sun's radiation, but partially blocking the infrared, long-wavelength radiation that's reflected from the surface of the earth.

After a century of burgeoning industrialisation, which produced heat-trapping gases in ever greater quantities, scientists began to realise just how powerful their effects might be. In the late 1950s, an American chemist Charles David Keeling persuaded the US Weather Bureau observatory near the peak of the Mauna Loa volcano on the 'Big Island' of Hawaii to use a technique he had developed for measuring concentrations of CO_2 in the atmosphere. The now famous 'Keeling Curve' shows mean levels of CO_2 to have risen steadily, from 316 ppm in 1959 to over 394 ppm in December 2012.[8]

We can look at these concentrations over a much longer time-frame by analysing air bubbles trapped in ice-cores and at the bottom of the oceans. Drilling at the Vostok ice core in Antarctica can retrieve, at a depth of 3,300 m, bubbles of air from the atmosphere of over 400,000 years ago.[9] The temperature record corroborates the findings from other palaeosciences of a 100,000-year cycle of warming and cooling over the past million years, ice ages alternating with interglacial periods of warmth. These cycles correlate with changes in the earth's orbit, and so the warmings and coolings are an example of natural 'orbital' forcing. They also correlate very closely with changes in levels of carbon dioxide and methane so that the best scientific hypothesis is that the orbital forcing is amplified first by the positive feedback loops from increased concentrations of heat-trapping gases, and then by the 'albedo effect', whereby melting of polar ice reduces the amount of heat reflected back into the atmosphere. As the authors of 'Climate and Atmospheric History' write in their abstract:

> Atmospheric concentrations of carbon dioxide and methane correlate well with Antarctic air-temperature throughout the record. Present-day atmospheric burdens of these two important greenhouse gases seem to have been unprecedented during the past 420,000 years.

We see considerable variation in levels of carbon dioxide over this long period – but nothing like the jump that kicked in after the Industrial Revolution. And if we look at more recent temperatures on the surface of the earth, as presented by the Goddard Institute of Space Studies at NASA, we see a steady increase since 1880, except for a levelling off between the 1940s and 1970s (Voiland, 2010). The last decade has brought mean temperatures to the highest levels ever recorded.

Let us now consider the steady rise in concentrations of atmospheric CO_2 charted in the Keeling Curve together with the increase in industrial emissions of CO_2 from fossil fuel combustion and cement production over the same period. (Land-use change and deforestation also contribute, but less than 10%.) We know that around half of the CO_2 we put into the atmosphere is absorbed by trees in spring and summer and also by the world's oceans; and if we assume that 57% of fossil-fuel emissions remain airborne, we get a perfect fit between the emissions curve and the CO_2 levels, which suggests that the increasing CO_2 levels is a direct result of human activities.[10] It is this remarkable fit that no climate sceptic has ever seen fit to address, since to dismiss it as mere coincidence would be absurd. And since John Tyndall's discovery of the heat-trapping properties of gases like CO_2 and methane has never been refuted but only corroborated, the obvious conclusion is that recent global warming is to a great extent anthropogenic.

There are, of course, many more factors involved in climate change than those mentioned above: the mechanisms behind climate forcing are complex, and the scientific understanding of climate sensitivity is far from complete. Nevertheless, the vast majority of climate scientists agree that global warming since the Industrial Revolution is being caused by human activities. There are, of course, climate sceptics, who disagree, but they are distinguished more by the shrillness of their rhetoric than the rigor of their science, and very few of them work at the best universities and institutes in climatology. A Stanford University study of the research of 1372 climate scientists showed that 97–98% of the scientists most actively publishing in the field of climate studies 'support the tenets of ACC [anthropogenic climate change] outlined by the Intergovernmental Panel on Climate Change', and also that 'the relative climate expertise and scientific prominence of the researchers unconvinced of ACC are substantially below that of the convinced researchers' (Anderegg, Prall, Harold, & Schneider, 2010, abstract).[11]

Although natural scientists can be contentious types, they also tend to be conservative and cautious about issuing bold public statements; and on top of that the best climate scientists are usually specialists, and so are reluctant to speak out about the general situation. A salient exception here is James Hansen, who has been Director of the Goddard Institute of Space Studies at NASA (the National Aeronautics and Space Administration) for over 30 years, and is also an adjunct professor in the Department of Earth and Environmental Sciences at Columbia University. Hansen has had a distinguished career in climatology and is a clear articulator of its findings: in 1988 he presented eloquent testimony to the US Congress about climate change, and has since then become more activist, which has led to his being arrested several times in recent years. His book *Storms of My Grandchildren* (2010) provides a clear and coherent overview of the perils of our warming the planet.[12]

Professional organisations of scientists are even more cautious than their individual members, and, in general, the more prestigious the organisation, the more carefully considered the statements it issues. Such professional bodies have an enormous stake in upholding their reputations and so take great pains to avoid saying anything that could possibly make them look stupid. Nevertheless, a recent one-year period saw the appearance of the following pronouncements:

The G8 + 5 Academies' joint statement: 'Climate change and the transformation of energy technologies for a low carbon future', May 2009 (signed by the National Academies of Sciences of Brazil, Canada, China, France, Germany, India, Italy, Japan, Mexico, Russia, South Africa, the United Kingdom and the United States).

President's Statement, US National Academy of Sciences, 'Avoiding Unmanageable Climate Change', October 2009 (Cicerone, 2009).

The presidents of 18 American scientific societies wrote to the US Senate urging action, and offering advice, on the ground that 'climate change is occurring, and rigorous scientific research demonstrates that the greenhouse gases emitted by human activities are the primary driver'.[13]

'Climate Change and the Integrity of Science', a 'Lead Letter', signed by 255 members of the US National Academy of Sciences, including 11 Nobel laureates, originally sent to the *Wall Street Journal* (Gleick et al., 2010).

The Royal Society (UK), 'Climate change: a summary of the science', September 2010.

An overwhelming consensus of the world's top climate scientists and their professional organisations acknowledges that climate change poses serious risks to the well-being of large numbers of human beings if we continue 'business as usual' in our burning of fossil fuels.[14]

The most important international body of scientists (and policy experts) in this area is the United Nations Intergovernmental Panel on Climate Change (IPCC), which has issued numerous reports since its inception in 1988. Even though these reports, based on reviews of the scientific literature by hundreds of scientists from all countries in the United Nations, tend to be very conservative, they have been loudly criticised by global warming deniers for containing occasional errors. Given the meticulous and responsible manner in which the IPCC operates, it is absurd to deny credence to their findings – which show more and more that climate change is largely human-caused and likely to have consequences that are catastrophic if we do not act to mitigate the warming.[15]

In November 2008, James Hansen and eight co-authors published a paper titled 'Target Atmospheric CO_2: Where Should Humanity Aim?' where they write:

> If humanity wishes to preserve a planet similar to that on which civilization developed and to which life on Earth is adapted, paleoclimate evidence and ongoing climate change suggest that CO_2 will need to be reduced from its current 385 [parts per million] to at most 350 ppm. (p. 217)

At the United Nations climate conference in Copenhagen the following year, 112 countries endorsed the 350 ppm target, mostly the poorer nations who are already suffering from the effects of global warming. But since they are relatively powerless in geopolitical terms, their endorsements mean almost nothing: the most advanced nations, with the exception of a number of countries in Europe, are intent on continuing with business as usual. We reached, and surpassed, a CO_2 concentration of 350 ppm in 1988, and the figure at the end of 2012 was over 394 ppm.

Among the uncertainties that the science of ACC necessarily involves, most do not detract from the big picture, but one is of crucial importance: concerning the effect of the aerosols produced by fossil fuel burning. The warming from the increased concentrations of CO_2 is offset – but we cannot tell by how much – by a cooling from the release of aerosols in burning fossil fuels. In other words, the cooling from particulate air pollution is masking the warming effect of increased carbon dioxide. Air pollution is already a huge public health issue worldwide, and as the rate of burning continues to increase, there will be more frequent calls to reduce air pollution because of its effect on public health.[16] In the worst case, as Hansen writes:

> If aerosols have been masking most of the greenhouse warming ... and humanity reduces particulate pollution by even half, the net climate forcing would double. *That increased forcing, combined with a continued greenhouse gas increase, might push the planet beyond tipping points with disastrous consequences* [emphasis added]. The current smaller net climate forcing already is causing a notable recession of mountain glaciers around the world, affecting freshwater availability, shifting climatic zones, increasing fires and flooding, promoting the loss of Arctic sea ice and vulnerable coral reefs, accelerating mass loss from the Greenland and Antarctic ice sheets with rising sea level, and putting pressures on many species, leading to a danger of mass extinctions. (2010, pp. 99–100)

What makes the prospects terrifying is that some of these factors – the loss of sea ice in the Arctic and ice sheets in Greenland and Antarctica, together with the release of vast amounts of frozen methane as tundra melts – can reach climate 'tipping points', thresholds beyond which there's no way to reverse the changes. Hansen calls what pushes the situation towards tipping points 'amplifying feedbacks', also known as positive feedback loops.

In the case of ice loss, it is the 'albedo effect', whereby more of the incoming radiation from the sun is reflected back into space by the whiteness of ice than the blueness of sea: the more the ice melts, the more heat from the sun the planet absorbs. With the tundra, as the planet warms, the permafrost thaws and releases organic carbon trapped in the ice. The tundra contains an estimated 1,600 gigatons (Gt) of carbon – more than twice the amount currently in the atmosphere (Tarnocai et al., 2009). Much of the carbon is released in the form of methane, a heat-trapping gas that is 25 times more effective than carbon dioxide. This release of CO_2 and methane in turn increases atmospheric warming, which means more melting – and so forth, in a vicious spiral. What is unsettling here is that the warming effects from thawing permafrost have never been taken into account by the assessment reports on climate change issued by the IPCC. The United Nations Environment Programme (UNEP, 2012) published a report in November 2012 drawing attention to the syndrome and recommending that the IPCC study the issue.

A third syndrome that may lead to a tipping point involves droughts in the Amazon Basin. Some climate models suggest that global warming may be giving rise to more frequent and severe droughts there, which in turn initiate 'positive feedbacks that could lead to widespread Amazon forest degradation or loss'. This would transform the Amazon Basin from a carbon sink into a huge *emitter* of CO_2. As the authors of a recent study in *Science* put it:

> The two recent Amazon droughts demonstrate a mechanism by which remaining intact tropical forests of South America can shift from buffering the increase in atmospheric carbon dioxide to accelerating it. (Lewis, Brando, Phillips, van der Heijden, & Nepstad, 2011, p. 554)

If we pass these tipping points, there's no way to prevent what the climate scientists call 'runaway global warming': even if we quickly stop burning fossils fuels, the feedback loops will spin the situation out of our control. As has happened before in the history of the earth's climate, the amplifying feedback that sets in after a tipping point is passed drives climate to an extreme of heat. Eventually the heat will evaporate the oceans and we end up with what Hansen calls 'the Venus Syndrome' – a fried planet with an atmosphere containing almost 97% carbon dioxide.[17]

There's a consensus in the climate sciences that we have to keep the planet's warming to under 2°C above pre-industrial levels, if we are to avoid exceedingly unpleasant consequences distributed unjustly among the world's various populations. Even though the UN Conference on Climate Change in Cancun in December 2010 was unable to achieve any binding agreements, the participating nations did manage a resolution 'to establish clear objectives for reducing human-generated greenhouse gas emissions over time to keep the global average temperature rise below two degrees' (Cancun Agreements). Although many climate scientists advise trying to keep the warming under 1.5°C rather than 2°C, the latest projections show that we are heading, in the absence of decisive action, to an increase by the year 2100 of 'about 4° to 6° above pre-industrial times' – where even the lower figure guarantees catastrophe (Global Carbon Project, 2012). A contemporary report from the International Energy Agency (IEA), 'Tracking Clean Energy Progress', says that

> achieving the transition [to the 2° scenario] is technically feasible, if timely and significant government policy action is taken, and a range of clean energy technologies are developed and deployed globally. (p. 5)

In December 2012, the executive director of the IEA, Maria van der Hoeven, issued the following warning to the participants of the UN Climate Conference at Doha:

> The need to rapidly transition to a more secure, sustainable global energy system is more urgent than ever. IEA analysis shows that achieving the internationally agreed climate goal of limiting warming to 2°C is becoming more difficult *and more expensive* with every passing year … . Yet even if the global temperature increase is limited to

> only 2°C, a warming planet may negatively affect our energy supply, demand and assets. In short, our energy security could be at risk. (para 1, emphasis added)

The problem is that the relevant governments show no interest in taking the 'timely and significant policy action' that would lead to a secure and sustainable global energy system – even though such action can be taken, given appropriate levels of investment. And so we will bequeath to our grandchildren a profoundly damaged world rife with chaos.

We are not dealing here with some vague and abstract hypothesis about how things could turn out: the changes that the climate models foresee – more frequent 'extreme weather events' for example – are already happening. In the United States, where denial of AGW is strongest and most widespread, two hurricanes just seven years apart, sisters Katrina (2005) and Sandy (2012), wreaked destruction to the tune of 150 billion dollars' worth of property damage. Climate scientists agree that it is premature to claim that global warming is the cause of the increase in extreme weather events we have seen over the past decade, but everything we know about the effects of higher temperatures on climate suggests that it is a major factor.

A rise in average temperatures means that the air can hold more moisture, which leads to rainstorms and flooding. Hurricanes and tropical cyclones may not get much more frequent with further global warming, but they are expected to become more intense and to bring heavier rainfall – which means greater destruction of people's livelihoods and homes (Knutson, 2008; Knutson et al., 2010). Warmer temperatures cause the oceans to expand, which leads to a rise in sea level, flooding of low-lying coasts, and salinisation of aquifers for drinking water. Although warming may improve growing conditions in some parts of the world for a time, it will disrupt agriculture elsewhere, promote desertification, expand the regions of the globe that are susceptible to tropical diseases, aggravate wars over resources, and thereby wreak prolonged mayhem. Climate models of what happens with global warming predict colder winters as well as more frequent droughts.

In view of this dismal scenario, where we visit a host of plagues upon innocent future generations, *what is to be done*? – Well, many people would say, and are saying: 'Nothing'. No need for any drastic changes in the way we consume energy.

CLIMATE SCEPTICISM

The rise to prominence of the climate sceptics and global warming deniers with their questioning of the scientific evidence for human-caused global warming is a bizarre phenomenon with a variety of sources. It became possible only because of the extreme complexity of the climate sciences and some strange changes that have taken place in the United States in recent decades. The problem is compounded by a number of factors ranging from the psychological, through the sociological, economic and religious, to the political and geopolitical and the field of transgenerational ethics. Let us begin by looking at the development of the climate scepticism business (and it is a lucrative business) in the United States.

A Brief History

In 2008, two American political scientists and a sociologist published a paper titled 'The organisation of denial: Conservative think tanks and environmental skepticism' (Jacques, Dunlap, & Freeman, 2008). Their investigation into conservative think tanks concerned with environmental issues discovered that 90% advocated environmental scepticism, and their analyses of over 140 'climate-sceptical' books, most of them published in the United States since 1992, revealed that over 92% of them were linked to conservative think tanks. Among such institutions, the most prominent are the George C. Marshall Institute, the American Enterprise Institute, the Heritage Foundation, the Cato Institute, the Competitive Enterprise Institute, and the Heartland Institute. Owing to the leaking of some internal documents from Heartland in February 2012, we now know what was long suspected: that it receives funds from fossil fuel concerns and one anonymous donor (of almost $9 million over four years recently) for its 'climate change' conference series, as well as plans for a consultant to develop a 'global warming curriculum' for schools – one that 'shows that the topic of climate change is controversial and uncertain'.[18]

An exemplary institution for the business of scepticism about science more broadly is the Marshall Institute, whose slogan on the masthead of its website is 'Science for Better Public Policy'. Three prominent physicists were involved in the Institute's founding in 1984: Robert Jastrow, William Nierenberg, and Frederick Seitz. One of the Institute's first major projects was to promote Ronald Reagan's 'Strategic Defense Initiative', better and more aptly known as the 'Star Wars' fantasy, and to defend its feasibility against the attacks from the majority of scientists – which eventually led to its demise. A cultural-historical analysis of the activities of the three Marshall Institute physicists shows them to have enjoyed enormous influence on politicians and policy makers, yet also to have held on to outmoded cultural values concerning science, technology and society, values that were threatened by on-going discoveries concerning the environment (Lahsen, 2008). As Clive Hamilton puts it, in the course of a discussion of the conservative activist Dixy Lee Ray: 'their anxiety over national sovereignty was matched by the disquiet they felt at environmentalism's destabilization of the idea of progress and mastery of nature' (2010, p. 100).[19]

Along with another physicist, Fred Singer, who is still active on behalf of contrarian concerns, the Marshall Institute 'trio' vigorously lobbied against a series of initiatives designed to protect public health and the natural environment. They questioned the links between smoking (and 'second hand smoke') and lung cancer, the contribution of coal and oil-fired power stations to the production of acid rain, the role of chlorofluorocarbons in ozone depletion, and more recently and energetically the human contribution to global warming. In each case, they dismissed the scientific consensus as 'bad science', or 'junk science', and advocated instead – with a confident zeal that would have amazed George Orwell – their own contrarian 'sound science' instead.

However, their activities consist not in doing scientific research of their own but rather in 'merchandising doubt', as Naomi Oreskes puts it, whose book with Erik M. Conway gives a meticulous account of this long and dismal tale of deceit (Oreskes & Conway, 2011; see also Hoggan, 2009). A banner at the foot of the Marshall Institute's home page reads, 'Responding to Oreskes and Conway's *Merchants of Doubt*': click on it and you're taken to a page with links to three responses to the book, authored by the President, the Director, and the CEO of the Institute.[20] Predictably, they dismiss the book as 'not history, [but] a one-sided polemic', but they make no attempt to respond to its arguments or refute the evidence that its authors provide. As a professor of history and science studies at the University of California, San Diego, Oreskes has great credibility and a huge professional stake in getting the facts right and making her arguments sound.

A rare 'insider' perspective on the Marshall Institute is provided by the author Matthew B. Crawford, who was appointed executive director in 2001, but resigned after five months. He had discovered the position was 'not to [his] taste' because:

> the trappings of scholarship were used to put a scientific cover on positions arrived at otherwise. These positions served various interests, ideological or material. For example, part of my job consisted of making arguments about global warming that just

> happened to coincide with the positions taken by the oil companies that funded the think tank. (2010, p. 82)

As Oreskes and Conway have pointed out, there *is* no 'disagreement' about the science of global warming, because that is not how the sciences work. They are a fundamentally collective enterprise that advances knowledge through the process of peer review. The Marshall Institute physicists had no expertise in the climate sciences, and instead of publishing research in peer-reviewed journals, they chose to disagree with the scientific consensus of experts in the forum of public opinion and policy making.

The hypothesis that the burning of fossil fuels is the major factor causing the planet to heat up is especially strong because of its fit with evidence from such a wide range of other sciences: with the physics of the selective absorptivity of greenhouse gases, and the findings of palaeogeography, palaeogeology, palaeobiology, palaeobotany, and above all with palaeoclimatology. This last branch of natural science correlates evidence of meteorological conditions with other processes on the earth record, and in this way can account for almost all periods of warming and cooling in the earth's history with only a few anomalies. While there are always uncertainties, the sceptics' claim that more research is needed, or that what the experts in climate science are telling us is wrong, is absurd – since it is inconceivable that everything the physical sciences have been telling us about the natural world since the seventeenth century could turn out to be false.

The grateful embrace of doubt about global warming by an indolent public is possible only on the basis of a misunderstanding of how the natural sciences work, a problem that is aggravated by the variety of the climate sciences and the complexity of their findings. This means that, since scientists usually specialise in a particular area of research, only a very few climatologists command an overview of the entire field that would allow them to be effective communicators of its findings to the public. On the other hand, the complexity of the climate sciences makes it possible for clever global warming deniers to seem at first blush as if they are making scientific sense, when in fact their presentations are full of misinterpretations, mistakes, misleading statistics and downright lies. But the surface of their presentations is so slick that it is hard for non-experts to discern the deception.

Two features of natural science are especially misunderstood: the role of *certainty* and the nature of *causality*. Climate sceptics say that the scientific evidence of AGW is not yet sufficient to warrant certainty. Politicians love to hear this, since it allows them in turn to say (as George W. Bush said throughout the six years of his presidency): 'We need to wait until the science of global warming is certain'. This is nonsense, because certainty is unattainable in the climate sciences. Meteorology, oceanography and climatology are not in a position to perform laboratory experiments to test their hypotheses. The laboratory would have to be the entire biosphere, where too many variables abound to permit controlled experiments.

Instead the climate sciences use statistics to calculate probabilities, and employ computer modelling to try to understand patterns of climatic change. When considering what action to take in the face of uncertainty, climate scientists (and scientists in many other fields) recommend following the 'Precautionary Principle'. This says that, in situations where scientific evidence gives us good reason to suppose that certain activities will prove harmful in the long run, the burden of proof should lie with those proposing to undertake or continue such activities, and that action should be taken to prevent or minimise harm even when the scientific evidence does not provide full certainty. At the UN 'Earth Summit' in 1992, representatives from 194 nations agreed that, 'In order to protect the environment, the precautionary approach shall be widely applied by States according to their capabilities' (United Nations Framework Convention on Climate Change, 1992, article 3, section 15).[21] So, to say that we should wait because the science is not yet certain is to miss the point completely. It would be like waiting for Godot, except that the stakes are higher.

As for causality, we tend to take examples from Newtonian physics as the paradigm, where a single cause brings about a single effect: the impact of the white billiard ball causes the red one to roll across the baize and into the pocket. But events in nature are rarely that simple: usually an event is brought about by multiple causes in dynamic interaction, as when a tropical storm results from a confluence of many weather conditions. The causality behind the urgent questions in epidemiology and public health are often of this nature: although we may be able to determine that the presence of a certain bacterium in the water supply is causing a certain disease, it is very hard to prove that industrial chemicals being discharged into a river are causing leukaemia and cancers in the local population. Similarly with cigarette smoking and lung cancer – a fact that allowed the tobacco companies to deny that smoking causes cancer, since strictly speaking we can only say that the greater the number of cigarettes smoked, the more likely the smoker is to develop the disease.

One often hears the objection, 'Carbon dioxide is only *correlated* with temperature, and there's no proof that higher concentrations of greenhouse gases actually *cause* rises in the earth's temperature'. Numerous websites devoted to denying anthropogenic climate change prominently display this falsehood on their front pages. What John Tyndall discovered about the selectively absorptive properties of carbon dioxide and other greenhouse gases has never been refuted, and it fits with everything else physics tells us about the interactions between solar radiation and gases in the atmosphere. Although there are other, minor factors also at work, there can be no reasonable doubt that these heat-trapping gases are major contributors to the rise in temperatures.

Psycho-Social Factors

Another problem with global warming which makes the sceptics' job easier is its *abstract* nature: it is a phenomenon whose dangers are not immediately manifest, where the risk is incremental and not directly perceptible by the senses. The rise in global temperatures is in any case very gradual, and accompanied by geographical fluctuations.

Human beings evolved over millions of years in hostile environments, and those of us who have survived have done so because our ancestors developed effective physical responses to imminent threats of harm. But the dangers from global warming are more distant and diffuse: indefinite in their timing, and uncertain in their effects on any particular individual. We tend to be lazy: we are creatures of habit and so dismiss information, or resist forces, that would oblige us to change our ways. And the habits that help us negotiate the difficulties of daily life are of no help when it comes to dealing with climate change. We also prefer effort to be closely followed by gratification, and so are reluctant in the case of global warming to forgo present pleasures for the sake of future generations – since most of these do not even exist yet.

The American Psychological Association, aware of the strength of psychological barriers to taking action on climate change, set up a task force to investigate the issue. Its report, 'Psychology and Global Climate Change' (2010), discusses several such barriers:

> Many … [people] are unaware of the problem, unsure of the facts or what to do, do not trust experts or believe their conclusions, think the problem is elsewhere, are fixed in their ways, believe that others should act, or believe that their actions will make no difference or are unimportant compared to those of others. (p. 8)

The report concludes by offering a number of research and policy recommendations in response to such problems.

Sociological studies suggest that cultural values and conformity to the views of one's social group – which condition 'what kind of person one is' – play a major role in determining what one makes of the evidence for ACC. Whether you identify yourself as an individualist or a communitarian, for example, makes a difference to the conclusions you draw from the evidence. Dan Kahan (2012) and others have pointed out that the 'costs' to the individual citizen of not agreeing with the scientific consensus on climate change are zero, while the costs of not conforming with the members of one's social group can be high. It is therefore quite rational under such circumstances to conform (Kahan, 2012, p. 255).[22]

Surprisingly, these studies show further that the more scientifically literate and numerate the individual, the more likely he or she is to become *more* polarised pro or con on the question of ACC. This suggests that cultural conformity with one's social group – together with 'the citizens' effective use of their knowledge and reasoning capacities to form risk perceptions that express their cultural commitments' – is a deeper and more powerful motivator than the common good, whether of the society or humanity as a whole.[23] This points up the shortcomings of any ethics that is focused on the smaller social group rather than the society, or the planet, more broadly.

Another factor here is the strange change that has taken place in the United States with respect to the status of natural scientists. Just 10 or 20 years ago, the general public trusted and respected its scientists, especially when they appeared on television wearing white lab coats. One reason for the change is that debates within the scientific community used to take place 'in private', as it were: the public would see only the generally agreed on results of scientific inquiries. But now the conservative think tanks have the finances and political connections to ensure that any missteps on the part of climate scientists are widely and shrilly reported in the media. The media have not served the public interest well in this area, as evidenced in the so-called 'Climategate fiasco'. On the basis of leaked emails from a number of climate scientists, a scandal was fomented in the press over alleged misconduct and skulduggery. Global warming deniers exploited this event shamelessly, encouraging the public to believe that most, if not all, climate scientists were liars and crooks. Subsequent independent investigations exonerated the scientists of any professional wrongdoing, but this was only very quietly reported in the media.[24]

Apart from the restricted sphere of their direct experience, the main way people 'know' about the world is through the mass media: newspapers and books, television, radio and the Internet. But the complexities of climate change do not lend themselves well to representation in such media, and if, as often happens, television affords equal opportunity to the climate sceptics and global warming deniers (in the interests of 'fair and balanced' reporting), this gives the false impression that the scientific community is more or less evenly divided on the issue – which, of course, it is not. What is more, many of the media can be 'bought' by parties with vested interests in preventing the bad news about fossil fuel burning from reaching the public. Some media have already been bought: witness the power of someone like Rupert Murdoch (in the news himself in 2012 because of the *News of the World* phone-hacking scandals) to influence public opinion.

The 'Climate Change and the Integrity of Science' letter mentioned earlier, signed by 255 members of the US National Academy of Sciences, was first sent to Murdoch's *Wall Street Journal*, which declined to publish it. Instead, the *WSJ* saw fit to run an 'Opinion' piece with the reassuring title, 'No Need to Panic about Global Warming: There's no compelling scientific argument for drastic action to "decarbonize" the world's economy'. The piece was signed by 16 scientists, none of whom is a prominent climate scientist and only three of whom appear to have anything to do with the climate sciences.[25] This publication unleashed a storm of criticism, led by a letter to the editor by 38 of the world's top climatologists who pointed out the authors' lack of relevant qualifications. Asking advice from that group about climate change would be like consulting your dentist about a heart condition:

> While accomplished in their own fields, most of these authors have no expertise in climate science. The few authors who have such expertise are known to have extreme views that are out of step with nearly every other climate expert.[26]

As several commentators have confirmed, the *WSJ* Opinion piece is full of factual errors, specious claims and misleading representations. An especially egregious misrepresentation concerns the eminent economist William Nordhaus.

Economics

In 'No Need to Panic' the "Wall Street Sixteen" wrote the following:

> A recent study of a wide variety of policy options by Yale economist William Nordhaus showed that nearly the highest benefit-to-cost ratio is achieved for a policy that allows 50 more years of economic growth unimpeded by greenhouse gas controls. (para 12)

And Professor Nordhaus's response:

> The piece completely misrepresented my work. My work has long taken the view that policies to slow global warming would have net economic benefits, in the trillion of dollars of present value. This is true going back to work in the early 1990s (MIT Press, Yale Press, Science, Proceedings of the National Academy of Science, among others). I have advocated a carbon tax for many years as the best way to attack the issue. (Revkin, 2012, para 13)

As often happens in such situations, the *Wall Street Journal* did not (as far as I know) correct the gross misrepresentation of Nordhaus's position, or any of the other errors the piece contained, and since the letter from the Climate Sciences Thirty-Eight does not mention Nordhaus, most readers will come away with completely the wrong impression of his position. He did, however, go on to deliver a comprehensive refutation of the sceptics in a long essay in the *New York Review of Books* (Nordhaus, 2012; see also Nordhaus, Cohen, Happer, & Lindzen, 2012).

Until recently, the environmental costs of doing business were largely ignored, since the natural world was thought to be vast enough to act as a bottomless 'sink' for the wastes produced by human industrial activity.[27] Sewage was discharged untreated into the sea, by-products of manufacturing processes were poured into rivers and lakes, and chemical pollutants and soot were spewed into the atmosphere – all on the assumption that the environment would absorb them well enough to forestall adverse effects on humans. But it soon became clear that this assumption was untenable, and that it makes sense to 'internalise' costs to the environment. If I operate a factory that discharges toxic waste into a nearby river, the damage to the environment and the neighbouring community is what economists call a 'negative externality', since it is borne by people and things that stand outside the economic transactions between me (as owner of the factory) and my customers. If instead I pay another company to remove the toxic waste on a regular basis, and incorporate the expense into the cost of what the factory produces, I am thereby 'internalising' the costs to the environment. This would be an example of the 'polluter pays' principle, which is widely regarded as fair enough – although companies often do not observe it if they think they can get away with it.

The 'Rio Declaration on Environment and Development', a document that arose from the UN 'Earth Summit' in 1992, contains an eminently sensible principle titled 'Internalization of Environmental Costs', which reads as follows:

> National authorities should endeavour to promote the internalization of environmental costs and the use of economic instruments, taking into account the approach that the

polluter should, in principle, bear the cost of pollution, with due regard to the public interest and without distorting international trade and investment. (The United Nations Conference on Environment and Development, 1992, Principle 16)

However, despite the fact that most of the consequences of global warming are against 'the public interest', the fossil fuel burning industries continue to assign their waste products, in the form of CO_2, to economic externalities. The world would be a very different place if economies included environmental costs (which include harm to humans) as internalities. A tax on carbon would be the obvious way to start, as Nordhaus and other economists have been suggesting for some time. Under the present circumstances, as the British economist Nicholas Stern put it in his report on *The Economics of Climate Change*: 'Climate change presents a unique challenge for economics: it is *the greatest example of market failure we have ever seen*.'[28] It is a failure because it is naïve to believe that we can continue to treat the atmosphere as 'an open sewer' into which we can pour our wastes without incurring crippling costs in the longer term.[29]

This is precisely what the 'merchants of doubt' examined by Oreskes and Conway refuse to acknowledge: what underlies and drives their contrarian resistance to scientific evidence is a fanatical adherence to 'free market fundamentalism', the belief that a laissez-faire capitalism that keeps markets free of government regulation is the only viable economic system.[30] It has proved viable for a while – though at great expense to non-capitalists and the natural environment – but its days are clearly numbered.

Two years after winning the Nobel Prize in Economics, Paul Krugman wrote a long essay for the *New York Times* titled 'Building a Green Economy', which he begins by remarking on the similarity between the debate over the science of climate change and the economic debate over the costs of continuing with business as usual versus taking action to avoid catastrophic economic consequences.

The casual reader might have the impression that there are real doubts about whether emissions can be reduced without inflicting severe damage on the economy. In fact, once you filter out the noise generated by special-interest groups, you discover that there is widespread agreement among environmental economists that a market-based program to deal with the threat of climate change – one that limits carbon emissions by putting a price on them – can achieve large results at modest, though not trivial, cost. (para 3)

Although Krugman opts for a different policy solution from Nordhaus's, he too argues that immediate economic action to reduce greenhouse-gas emissions is crucial, and would cost the world economy little by comparison with the price we pay if we delay. Comparing the costs of reducing emissions and mitigating greater damage later: 'Climate change will lower gross global product 5 percent, stopping it will cost 2 percent' (para 60). Krugman acknowledges the extreme complexity of the issue, which renders all predictions uncertain – but argues that in any case the significant risk of utter catastrophe must override simple cost-benefit calculations and requires strong climate policy to reduce emissions.

So why are the oil, gas and coal industries – let us call them 'the fossil fuel concerns' since they are going the way of the dinosaurs – so vehement in their opposition to the news about AGW and so wilfully blind to the arguments of the environmental economists? It was evident decades ago that the rational way for these industries to serve the public interest while at the same time maximising profits would have been – in view of pollution problems and the finite supply of fossil fuels – to reconfigure themselves as energy suppliers rather than fossil fuel burners. But they are obliged to maintain their 'see no evil' stance because profits have to come before the public interest, and their investments in fossil fuels are too huge. As Krugman remarks, public policy is obliged to take a far longer view of the situation than private markets.

On one hand, there are what business-people call the 'sunk costs' of the current fossil fuel infrastructure (from oil rigs in the Gulf of Mexico to pipelines and supertankers, coal mines and power plants) which have been estimated at over $10 trillion worldwide – a sum of money beyond the capacity of most of us to even imagine. That infrastructure is going to be around for up to 50 years before the capital costs will be paid off, and so, as Bill McKibben bluntly puts it, 'If we shut that infrastructure down early, merely to prevent ourselves from frying the planet, someone will have to eat those costs' (2011, p. 55).

On the other hand, there is the prospect of 'stranded assets' among the resources the fossil fuel concerns have their hands on but have not yet put on the market, and this poses a far larger financial problem. According to a report by the Carbon Tracker Initiative in London, the currently known fossil fuel reserves on the planet – listed on public company balance sheets and reports by the state-owned concerns in countries like China, Russia, Iran, and Venezuela – contain a total of 2,795 Gt of carbon dioxide: 65% from coal, 22% from oil, 13% from gas (Leaton, n.d.). The good news, for the fossil fuel concerns, is that these reserves are worth an estimated $27 trillion (Fullerton, n.d.). The bad news, for the rest of the human race, is that if these resources are sold and used, the planet will become uninhabitable.

Now consider this in the context of the scientific consensus that, if average global temperatures rise more than 2°C from pre-industrial levels, the effects on tropical disease vectors, desertification and drought, sea-level rise and floods, freshwater supplies and food production, will be catastrophic. (Many experts think that an increase of even 1.5°C would have disastrous consequences.) We have already seen an increase of 0.6°C, and are beginning to experience its effects in the form of the 'extreme weather events' that the climate models lead us to expect.

There were high hopes that the conference convened by the United Nations Framework Convention on Climate Change in Copenhagen in 2009 would finally, with the expiration of the Kyoto Protocol imminent, lead to some meaningful and binding agreement to slow global warming, but, mainly owing to lack of cooperation on the part of the United States and China, it did not. If there was a high point anywhere in those depressing proceedings, it was the acknowledgment in the (non-binding) 'Copenhagen Accord' of the scientific view that average temperatures should be kept below 2°C. Those nations that are already experiencing the adverse effects of global warming, the Alliance of Small Island States and the least developed countries, recommended for good reason a limit of 1.5°C. The conference as a whole agreed to review in four years the question of the limit to aim for.

In 2010, the United Nations Environment Programme (UNEP, 2010) published a document, 'The Emissions Gap Report', showing that, even if nations followed through on their (non-binding) pledges, it would be unlikely that even the 2°C increase could be avoided (UNEP, 2010). Nevertheless, the follow-up to Copenhagen, the UN Conference on Climate Change in Cancun in December 2010, was unable to achieve any binding agreements, only a resolution 'to establish clear objectives for reducing human-generated greenhouse gas emissions over time to keep the global average temperature rise below two degrees' (Cancun Agreements, 2010, first objective).

Research by the Potsdam Institute for Climate Impact Research in Germany, the foremost institute of its kind in the world, has estimated the upper limit to CO_2 emissions if we are to reduce to 20% the chances of exceeding a 2°C warming. The resulting 'carbon budget' for the years 2000–2050, the amount we can afford to burn without risking catastrophe, is 886 Gt (Meinshausen et al., 2009). Since we have already emitted, through fossil fuel burning and changes in land use, over 320 Gt in the century's first decade, this leaves a budget of around 565 Gt for the years up to 2050 (Leaton, n.d., p. 6). The IEA, in its *World Energy Outlook* for 2011, points out that four-fifths of the carbon budget to 2035 are already locked in place by capital investment in existing infrastructure (power plants, factories, etc.), so that the 2°C warming is becoming ever harder to avoid – and a 6°C warming is the likely outcome if we continue with business as usual (International Energy Agency and Organisation for Economic Co-operation and Development, 2011).

Bill McKibben recently laid out many of these considerations with admirable clarity and concision in an essay titled 'Global Warming's Terrifying New Math' in *Rolling Stone* (McKibben, 2012). He points out that since the known fossil fuel reserves (2,795 Gt) are five times the amount that the climate scientists say is safe to burn (565 Gt), around 80% of them will have to stay in the ground – which would mean a loss of some $20 trillion to the fossil fuel concerns. Add to that whatever percentage of the $10 trillion in existing infrastructure has not been paid off yet, and it is no wonder they are fighting so hard to persuade people that global warming is not our (or their) fault.

The second part of this essay considers the major obstacles to taking action on global warming and evaluates various proposed solutions.

NOTES

1. I would like to thank John Barry of Queens University Belfast for helpful comments on an earlier draft of this paper. I am aware of the terminological shift from talking about 'global warming' to discussions of 'climate change', but will use both terms in what follows, as well as the abbreviations for the human-caused versions (AGW for anthropogenic global warming and ACC for anthropogenic climate change). Though the changes brought about by the warming include colder winters and other extremes of climate, the warming remains the fundamental reality.
2. See, for example, Gallup and Newport (2010).
3. See, for example, Kasser (2003), Graham (2011, 2012), and Bok (2011).
4. See Dyer (2011) for an array of depressing facts and possible scenarios.
5. Atiq Rahman, executive director, Bangladesh Centre for Advanced Studies, cited in Dyer (2011).
6. The concentrations of CO_2 are measured and calculated at the Earth Systems Research Laboratory in Hawaii (Tans & Keeling, n.d.).
7. A picture of this magnificent device can be seen at http://doc-snow.hubpages.com/hub/Global-Warming-Science-In-The-Age-Of-Queen-Victoria.
8. See the website of the CO_2 program at the Scripps Institution of Oceanography at the University of California San Diego: http://scrippsco2.ucsd.edu/graphics_gallery/mauna_loa_record/mauna_loa_record_-_color.html; accessed on 23 August 2012. For the latest figures, see http://co2now.org/; accessed on 12 January 2013.

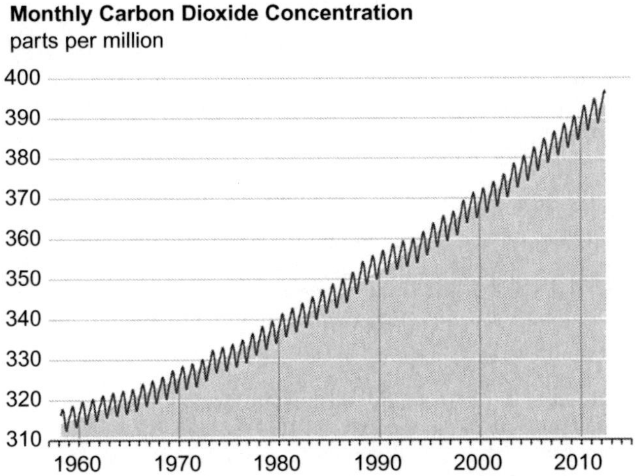

9. See Petit et al. (1999). See the chart of temperature and levels of carbon dioxide and methane in Chapter 2 of Intergovernmental Panel on Climate Change (2001).

10. Black (zigzag) curve: Monthly average atmospheric carbon dioxide concentration versus time at Mauna Loa Observatory, Hawaii (20°N, 156°W) where CO_2 concentration is in parts per million in the mole fraction (ppm). The curve is a fit to the data based on a stiff spline plus a 4 harmonic fit to the seasonal cycle with a linear gain factor. Red curve: Fossil fuel trend of a fixed fraction (57%) of the cumulative industrial emissions of CO_2 from fossil fuel combustion and cement production. This fraction was calculated from a least squares fit of the fossil fuel trend to the observation record. Data from Scripps CO_2 Program. Last Updated October 2009 (http://scrippsco2.ucsd.edu/program_history/keeling_curve_lessons_3.html).

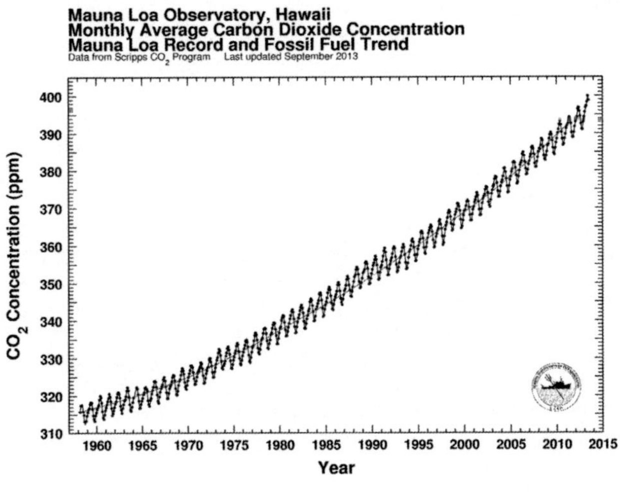

11. See also the earlier analysis by Oreskes (2004).

12. Hansen et al. (2008), *Storms of My Grandchildren: The Truth about the Coming Climate Catastrophe and Our Last Chance to Save Humanity*. The wordy subtitle sounds more like the publisher than Hansen, but the body of the text is sober in its elucidation of the scientific issues and comprehensive documentation of the dangers. For some literature by authors whose work is well grounded in up-to-date evidence from the climate sciences, see Kolbert (2006); Monbiot (2007); McKibben (2011); Hamilton (2010); and Hertsgaard (2012).

13. The American Association for the Advancement of Science, American Chemical Society, American Geophysical Union, American Institute of Biological Sciences, American Meteorological Society, American Society of Agronomy, American Society of Plant Biologists, American Statistical Association, Association of Ecosystem Research Centers, Botanical Society of America, Crop Science Society of America, Ecological Society of America, Natural Science Collections Alliance, Organization of Biological Field Stations, Society for Industrial and Applied Mathematics, Society of Systematic Biologists, Soil Science Society of America, University Corporation for Atmospheric Research; Letter available for viewing at: http://www.ucsusa.org/assets/documents/ssi/climate-change-statement-from.pdf, accessed 1 September 2012.

14. For a concise account, see Rockström et al. (28 distinguished co-authors) (2009).

15. For an excellent account of the IPCC, see Bolin (2008). See also footnote 29* in *The Politics of Global Warming (2): Two Obstacles to Circumvent* for an example of the IPCC's modus operandi.

16. Hansen (2010). A comprehensive study published by *The Lancet* estimates that pollution from cars caused over 2 million premature deaths in east and south Asia during 2010 (Lim et al., 2012). According to a report by the Organisation for Economic Co-operation and Development, we can expect 3.6 million premature deaths from ground-level ozone pollution alone by 2050 (OECD, 2012).

17. Hansen, 2010, Chapter 10, 'The Venus Syndrome'.

18. Hickman (2012), the documents, fraudulently obtained, are posted at Demelle, 2012. The institute has claimed that the 'Climate Strategy' memo is faked, but has provided no evidence.

19. See also his insightful account of climate change denial, pp. 95–133.

20. The Marshall Institute – Responding to Oreskes and Conway's Merchants of Doubt (n.d.).

21. Section 3 reads: "The Parties should take precautionary measures to anticipate, prevent or minimize the causes of climate change and mitigate its adverse effects. *Where there are threats of serious or irreversible damage, lack of full scientific certainty should not be used as a reason for postponing such measures* … taking into account that policies and measures to deal with climate change should be cost-effective so as to ensure global benefits at the lowest possible cost." (Emphasis added)

22. "So, if the cost of having a view of climate change that does not conform with the scientific consensus is zero, and the cost of having a view that is at odds with members of one's cultural community can be high, what is a rational person to do? In that situation, it is perfectly sensible for individuals to be guided by modes of reasoning that connect their beliefs to ones that predominate in their group. Even people of modest scientific literacy will pick up relevant cues. Those who know

more and who can reason more analytically will do a still better job, even if their group is wrong on the science."

23. Kahan et al. (2011). See also the section titled 'Values determine beliefs' in Hamilton (2010, 107ff).

24. The eight major investigations covered by secondary sources include the following: The House of Commons Science and Technology Committee (United Kingdom); Independent Climate Change Review (United Kingdom); International Science Assessment Panel (United Kingdom); Pennsylvania State University first panel and second panel (United States); United States Environmental Protection Agency (United States); Department of Commerce (United States); National Science Foundation (United States). Listed in Climate Research Unit email controversy, 2013.

25. Claude Allegre et al. (2012), "No Need to Panic about Global Warming."

26. The text of the letter and list of signatories can be read at Johnson (n.d.).

27. For an excellent account of the economic issues involved here, as well as the politics associated with them, see Barry (2012).

28. Stern has recently expressed regret that in fact the situation is far worse than he realised when writing that book: Stewart and Elliott (2013).

29. Peter Singer (2004) makes the analogy between emitting greenhouse gases and pouring our wastes into a huge sink in *One World: The Ethics of Globalization,* 27ff. Al Gore (2010) used the term 'open sewer' in an op-ed piece, 'We can't wish away climate change', in the *New York Times.*

30. Oreskes and Conway (2011, pp. 248–55). The term 'free market fundamentalism' comes from Soros (1998).

REFERENCES

Allegre, C., et al. (2012). No need to panic about climate change. *Wall Street Journal.* Retrieved from http://online.wsj.com/article/SB10001424052970204301404577171531838421366.html. Accessed on April 22, 2013.

Anderegg, W. R. L., Prall, J. W., Harold, J., & Schneider, S. H. (2010). Expert credibility in climate change. *Proceedings of the National Academy of Sciences.* doi:10.1073/pnas.1003187107

APA Task Force on the Interface Between Psychology and Global Climate Change. (2010). *Psychology and global climate change: Addressing a multi-faceted phenomenon and set of challenges.* Retrieved from http://www.apa.org/science/about/publications/climate-change.aspx. Accessed on April 22, 2013.

Barry, J. (2012). *The politics of actually existing unsustainability: Human flourishing in a climate-changed, carbon constrained world.* New York, NY: Oxford University Press.

Bok, D. (2011). *The politics of happiness: What government can learn from the new research on well-being.* Princeton, NJ: Princeton University Press.

Bolin, B. (2008). *A history of the science and politics of climate change: The role of the Intergovernmental Panel on Climate Change.* New York, NY: Cambridge University Press.

Cancun Agreements: main objectives of the agreements. (2010). *UNFCCC.int.* Retrieved from http://cancun.unfccc.int/cancun-agreements/main-objectives-of-the-agreements/#c33. Accessed on April 22, 2013.

Christianson, G. E. (1999). *Greenhouse: The 200-year story of global warming*. Vancouver: Greystone Books.

Cicerone, R. J. (2009). Address by Ralph J Cicerone. *National Academy of Sciences*. Retrieved from http://www.nasonline.org/about-nas/leadership/president/climate-change-october-2009.html. Accessed on April 23, 2013.

Climatic Research Unit email controversy. (2013). *Wikipedia, the free encyclopedia*. Retrieved from http://en.wikipedia.org/w/index.php?title=Climatic_Research_Unit_email_controversy&oldid=550235045

Demelle, B. (2012). Heartland Institute Exposed: Internal documents unmask heart of climate denial machine. *DeSmogBlog*. Retrieved from http://www.desmogblog.com/heartland-institute-exposed-internal-documents-unmask-heart-climate-denial-machine

Dyer, G. (2011). *Climate wars: The fight for survival as the world overheats*. London: Oneworld Publications.

Fourier, J. (1824). Remarques générales sur les températures du globe terrestre et des espaces planétaires. *Annales de Chimie et de Physique*, *27*, 136–167.

Fullerton, J. (n.d.). *The big choice. The future of finance*. Capital Institute. Retrieved from http://capitalinstitute.org/blog/big-choice-0

Gallup, & Newport, F. (2010). Americans' global warming concerns continue to drop. *Gallup.com*. Retrieved from http://www.gallup.com/poll/126560/Americans-Global-Warming-Concerns-Continue-Drop.aspx

Gleick, P. H., et al. (2010). Climate change and the integrity of science. *Science*, *328*(5979), 689–690. doi:10.1126/science.328.5979.689

Global Carbon Project. (2012). *Carbon budget and trends 2012*. Retrieved from www.globalcarbonproject.org/carbonbudget. Accessed on April 23, 2013.

Gore, A. (2010). We can't wish away climate change. *The New York Times*. Retrieved from http://www.nytimes.com/2010/02/28/opinion/28gore.html

Graham, C. (2011). *The pursuit of happiness: An economy of well-being*. Washington, DC: Brookings Institution Press.

Graham, C. (2012). *Happiness around the world: The paradox of happy peasants and miserable millionaires*. New York, NY: Oxford University Press.

Hamilton, C. (2010). *Requiem for a species: Why we resist the truth about climate change*. London: Routledge.

Hansen, J. (2010). *Storms of my grandchildren: The truth about the coming climate catastrophe and our last chance to save humanity*. New York, NY: Bloomsbury Press.

Hansen, J., et al. (2008). Target atmospheric CO_2: Where should humanity aim? *The Open Atmospheric Science Journal*, *2*(1), 217–231. doi:10.2174/1874282300802010217

Hertsgaard, M. (2012). *Hot: Living through the next fifty years on earth*. Boston, MA: Mariner Books.

Hickman, L. (n.d.). Leaked Heartland Institute documents pull back curtain on climate scepticism. *The Guardian*. Retrieved from http://www.guardian.co.uk/environment/blog/2012/feb/15/leaked-heartland-institute-documents-climate-scepticism

Hoggan, J. (2009). *Climate cover-up: The crusade to deny global warming*. Vancouver: Greystone Books.

Intergovernmental Panel on Climate Change. (2001). *Climate change 2001: The scientific basis: Contribution of Working Group I to the third assessment report of the Intergovernmental Panel on Climate Change*. New York, NY: Cambridge University Press.

International Energy Agency, & Organisation for Economic Co-operation and Development. (2011). *World energy outlook 2011*. Paris: IEA, International Energy Agency and OECD.

Jacques, P. J., Dunlap, R. E., & Freeman, M. (2008). The organisation of denial: Conservative think tanks and environmental scepticism. *Environmental Politics*, *17*(3), 349–385. doi:10.1080/09644010802055576

Johnson, B. (n.d.). Climate scientists rebuke Rupert Murdoch: WSJ Denier op-ed like "Dentists Practicing Cardiology." Retrieved from http://thinkprogress.org/climate/2012/02/01/416078/climate-scientists-rebuke-rupert-murdoch-wsj-denier-op-ed-like-dentists-practicing-cardiology/?mobile = nc. Accessed on April 22, 2013.

Kahan, D. (2012). Why we are poles apart on climate change. *Nature*, *488*(7411), 255–255. doi:10.1038/488255a

Kahan, D., et al. (2011). *The tragedy of the risk-perception commons: Culture conflict, rationality conflict, and climate change*. SSRN Scholarly Paper No. ID 1871503, Social Science Research Network, Rochester, NY. Retrieved from http://papers.ssrn.com/abstract = 1871503

Kasser, T. (2003). *The high price of materialism*. Cambridge, MA: MIT Press.

Knutson, T. R., et al. (2010). Tropical cyclones and climate change. *Nature Geoscience*, *3*(3), 157–163. doi:10.1038/ngeo779

Knutson, T. R. (2013). Global warming and hurricanes: An overview of current research results. *Geophysical Fluid Dynamics Laboratory*. Retrieved from http://www.gfdl.noaa.gov/global-warming-and-hurricanes. Accessed on May 1, 2013.

Kolbert, E. (2006). *Field notes from a catastrophe: Man, nature, and climate change*. New York, NY: Bloomsbury Press.

Lahsen, M. (2008). Experiences of modernity in the greenhouse: A cultural analysis of a physicist "trio" supporting the backlash against global warming. *Global Environmental Change*, *18*(1), 204–219. doi:10.1016/j.gloenvcha.2007.10.001

Leaton, J. (n.d.). Unburnable carbon: Are the world's financial markets carrying a carbon bubble? *Carbon Tracker*. Retrieved from http://www.carbontracker.org/carbonbubble. Accessed on April 22, 2013.

Lewis, S., et al. (2011). The 2010 Amazon drought. *Science*, *331*(6017), 554–554. doi:10.1126/science.1200807

Lim, S. S., et al. (2012). A comparative risk assessment of burden of disease and injury attributable to 67 risk factors and risk factor clusters in 21 regions, 1990–2010: A systematic analysis for the Global Burden of Disease Study 2010. *The Lancet*, *380*(9859), 2224–2260. doi:10.1016/S0140-6736(12)61766-8

McKibben, B. (2011). *Earth: Making a life on a tough new planet*. New York, NY: St. Martin's Griffin.

McKibben, B. (2012, July 19). Global warming's terrifying new math | Politics News | Rolling Stone. *rollingstone.com*. Retrieved from http://www.rollingstone.com/politics/news/global-warmings-terrifying-new-math-20120719. Accessed on April 22, 2013.

Meinshausen, M., et al. (2009). Greenhouse-gas emission targets for limiting global warming to 2℃. *Nature*, *458*(7242), 1158–1162. doi:10.1038/nature08017

Monbiot, G. (2007). *Heat: How to stop the planet burning*. London: Penguin Books.

No Need to Panic About Global Warming. (2012). *Wall Street Journal*, January 12. Retrieved from http://online.wsj.com/article/SB10001424052970204301404577171531838421366.html

Nordhaus, W. D. (2012). Why the global warming skeptics are wrong. *The New York Review of Books*. Retrieved from http://www.nybooks.com/articles/archives/2012/mar/22/why-global-warming-skeptics-are-wrong/

Nordhaus, W. D. et al. (2012). In the climate casino: An exchange. *The New York Review of Books*. Retrieved from http://www.nybooks.com/articles/archives/2012/apr/26/climate-casino-exchange/

OECD. (2012). *OECD environmental outlook to 2050: The consequences of inaction*. Paris: OECD.

Oreskes, N. (2004). The scientific consensus on climate change. *Science*, *306*(5702), 1686–1686. doi:10.1126/science.1103618

Oreskes, N., & Conway, E. (2011). *Merchants of doubt: How a handful of scientists obscured the truth on issues from tobacco smoke to global warming*. New York, NY: Bloomsbury Press.

Peters, G. P., et al. (2011). Rapid growth in CO_2 emissions after the 2008–2009 global financial crisis. *Nature Climate Change*, *2*(1), 2–4. doi:10.1038/nclimate1332

Petit, J. R., et al. (1999). Climate and atmospheric history of the past 420,000 years from the Vostok ice core, Antarctica. *Nature*, *399*(6735), 429–436. doi:10.1038/20859

Revkin, A. C. (2012). Scientists challenging climate science appear to flunk climate economics. *Dot Earth Blog*. Retrieved from http://dotearth.blogs.nytimes.com/2012/01/30/scientists-challenging-climate-science-appear-to-flunk-climate-economics/

Rockström, J., et al. (2009). A safe operating space for humanity. *Nature*, *461*(7263), 472–475. doi:10.1038/461472a

Singer, P. (2004). *One world: The ethics of globalization*. New Haven, CT: Yale University Press.

Soros, G. (1998). *The crisis of global capitalism: Open society endangered*. New York, NY: Public Affairs.

Stewart, H., & Elliott, L. (2013). Nicholas Stern: "I got it wrong on climate change – It's far, far worse." *The Guardian*. Retrieved from http://www.guardian.co.uk/environment/2013/jan/27/nicholas-stern-climate-change-davos

Tans, P., & Keeling, R. (n.d.). *Trends in carbon dioxide*. Retrieved from http://www.esrl.noaa.gov/gmd/ccgg/trends/. Accessed on April 23, 2013.

Tarnocai, C., et al. (2009). Soil organic carbon pools in the northern circumpolar permafrost region. *Global Biogeochemical Cycles*, *23*(2). doi:10.1029/2008GB003327

The Marshall Institute. (n.d.). Responding to Oreskes and Conway's merchants of doubt. Retrieved from http://www.marshall.org/article.php?id = 1079. Accessed on April 22, 2013.

U.S. Department of Defense. (2010). Quadrennial defense review. *Defense.gov*. Retrieved April 23, 2013 from http://www.defense.gov/qdr/

United Nations Conference on Environment and Development. (1992). *Rio declaration on environment and development*. Retrieved from http://www.unep.org/Documents.Multilingual/Default.asp?documentid = 78&articleid = 1163. Accessed on April 22, 2013.

United Nations Environment Programme. (2010). The Emissions Gap Report. Retrieved from http://www.unep.org/publications/ebooks/emissionsgapreport/index.asp. Accessed on April 22, 2013.

United Nations Environment Programme. (2012). *Policy implications of warming permafrost*. Nairobi, Kenya: United Nations Environment Programme.

United Nations Framework Convention on Climate Change. (1992). Retrieved from http://unfccc.int/home/items/6078.php?q = 1992&cx = 009772925632828311246%3Agjvsnghto1u&ie = UTF-8&sa = . Accessed on April 22, 2013.

Vidal, J. (2013). Large rise in CO_2 emissions sounds climate change alarm. *The Guardian*. Retrieved from http://www.guardian.co.uk/environment/2013/mar/08/hawaii-climate-change-second-greatest-annual-rise-emissions?INTCMP = SRCH

Voiland, A. (2010). NASA – 2009: Second warmest year on record; end of warmest decade. *NASA*. Feature. Retrieved from http://www.nasa.gov/topics/earth/features/temp-analysis-2009.html. Accessed on May 1, 2013.

THE POLITICS OF GLOBAL WARMING (2): TWO OBSTACLES TO CIRCUMVENT

Graham Parkes

ABSTRACT

Purpose – *The purpose of this chapter is twofold. First, to show how the financial power of the fossil fuel industries and the prevalence of religious ideology in Congress are the two major obstacles preventing the U.S. government from taking action to slow down global warming. Then to evaluate various approaches to 'satisfying our energy needs', by showing a crucial dynamic behind our insatiable drive to consume energy, and to propose some ways of circumventing the current obstacles.*

Methodology/approach – *The approach is through a comprehensive study of the relevant evidence and academic literature, interwoven with philosophical reflections on their significance.*

Findings – *The findings are as follows: a major root of the current problem is the dysfunctional political system in the United States, which is corrupted by vast infusions of money from the fossil fuel industries and the dogmatic religious beliefs of Republicans in key positions on Congressional committees.*

Environmental Philosophy: The Art of Life in a World of Limits
Advances in Sustainability and Environmental Justice, Volume 13, 81–109

ISSN: 2051-5030/doi:10.1108/S2051-5030(2013)0000013008

Social implications – *The implications are several. The proposed technological solutions to the 'energy problem' – nuclear power, carbon sequestration, fracking for natural gas and geo-engineering – only address the symptoms and ignore the dynamic that underlies them, exemplified in the story of Prometheus. If we continue to be driven by the Promethean spirit, we risk being subject to excruciating punishment as a result. The solution to our problems is a transition to clean and renewable sources of energy, accompanied by the kind of reduction in material desires that evidently makes for lives that are more fulfilled.*

Originality/value – *The value of the philosophical perspective on this topic is that it highlights questions of value that otherwise remain inexplicit.*

Keywords: Fossil fuel industry; politics; Prometheus; religious right; renewable energy.

TWO IMMEDIATE OBSTACLES

Politics in the United States

The most powerful factor in all of this is the way profits direct politics. Plato, in his *Republic*, sensibly disconnected the ruling of the polis, reign over the political, from monetary gain and material profit. In order to show what is the best kind of ruler, Plato has Socrates introduce the analogy of the doctor, whose concern is the advantage of the patient rather than his own. To this extent, one must distinguish doctors, as practitioners of an indispensable art, from 'businessmen' or 'money-makers'. In the case of the medical practitioner, it benefits society to disconnect personal gain from professional activity. If the health service pays me a fixed salary for being a surgeon, regardless of how many procedures I perform, I can decide whether treatment is called for purely out of consideration for the patient. Whether I operate or not, prescribe medicine or not, or do any kind of treatment, has no effect on my earnings. The same must hold for the ideal ruler:

> There isn't ever anyone who [responsibly] holds any position of ruler … who considers or commands his own advantage rather than that of what is ruled and of which he himself is the craftsman. (Book 1, 341c–342e)

The best rulers, according to Socrates, are those – strangely – who are reluctant to rule: after all, there's nothing in it for them personally, since genuine ruling has to be directed by the advantage of those who are ruled. The rulers are of a type that is not interested in either the profit or the honour that might come from ruling: they consent to rule only because there is no one better or more qualified, and they are reluctant to be ruled by those less capable than themselves. Socrates requires that the rulers possess no private property except for the necessities, which will be provided as a 'salary' by other citizens, in moderate amounts that result in neither surplus nor lack.[1] This arrangement ensures that the prospect of personal gain cannot possibly enter into the decisions the rulers make concerning what is best for the community as a whole.

When we turn from this ideal republic to the federal republic of the United States of America, we find a situation that could hardly be farther from what Plato envisaged. There is no place like the United States for profits directing politics: following decades of efforts by honourable politicians to reduce the impact of money on politics through various kinds of campaign reform, the shameful decision of the Supreme Court in the *Citizens United* case in 2010 basically affirmed the rights of the rich to buy the politicians and the policies they want. The 'Super PACs' and other monstrosities ushered in by the Supreme Court's decision are turning the country into a full-on plutocracy, where the 1% can rule the other 99 as they see fit. A report released by the Global Commission on Elections, Democracy and Security (a distinguished international body headed by Kofi Annan) is severely critical of this situation, observing that a series of court decisions culminating in Citizens United has 'undermined political equality, weakened transparency of the electoral process and shaken citizen confidence in America's political institutions and elections' (Global Commission on Elections, Democracy & Security, 2012).

The main obstacle to international agreements to tackle global warming is the US government, which refuses to take appropriate action – thereby providing nations like China and India with a justifiable pretext for likewise refraining from action.[2] The major factor here is the fossil fuel industry, which has a huge vested interest in continuing business as usual, and increasingly has the political power to ensure that it can. Governments in the United Kingdom and the United States have been giving generous subsidies to those industries for decades, which are in turn using their vast profits to maintain the political clout to keep those subsidies coming. According to the International Energy Agency (IEA), US taxpayers in 2008 contributed to subsidies for fossil fuel concerns – some of the most lucrative businesses on the planet – to the tune of a staggering $557 billion.[3]

All this in spite of the fact that in 2009 the leaders of the G20 nations committed themselves to phasing out subsidies that 'encourage wasteful consumption, reduce our energy security, impede investment in clean energy sources and undermine efforts to deal with the threat of climate change'. Analyses by the IEA and the OECD (Organisation for Economic Co-operation and Development) predict that removing inefficient subsidies, in conjunction with programmes that assist poor households and displaced energy workers, 'would raise national revenues and reduce greenhouse-gas emissions' (IEA and OECD, 2011). Nevertheless, in March 2012, the British government awarded the oil industry an additional £3 billion in tax breaks in order to open up an area to the west of the Shetland Islands for drilling new wells (Macalister, 2012). Since British Petroleum is one of the beneficiaries, one can only hope that the rigs they build work better than the Deepwater Horizon rig in the Gulf of Mexico, where an explosion in 2010 killed 11 workers and spewed some 5 million barrels of crude oil into the ocean before it was capped.

Since they stand to lose such a fortune, the fossil fuel concerns not only fund conservative think-tanks but also channel money to politicians in the U.S. Congress to get them to oppose any legislation that threatens to render their $10 trillion infrastructure investment redundant. Koch Industries, an oil and chemicals conglomerate that is the second largest private company in the United States, is one of the biggest spenders.[4] Since the beginning of the 2006 election cycle, the Koch Political Action Committee spent more on contributions to federal candidates ($2.5 million) than any other fossil fuel concern PAC (ExxonMobil spent $1.7 million). And between 2006 and 2009, they paid over $37.9 million for direct lobbying on issues in the oil and gas industries (Greenpeace, n.d., p. 8). In addition, between 2005 and 2008, foundations supported by Koch Industries donated $24.9 million to climate-sceptical organisations (far outspending ExxonMobil at $8.9 million).[5] Many of the recipients' names are familiar: the Mercatus Center, the Americans for Prosperity Foundation, the Heritage Foundation, the Cato Institute, the Marshall Institute, and the Heartland Institute. Since Koch Industries are only one among many big donors, it is no wonder that opposing reductions in CO_2 emissions from fossil fuel burning can be such a lucrative line of business.

In his carefully documented study of the outrages perpetrated against the sciences by the Republican Party, especially during the presidencies of Nixon, Reagan and the Bushes, Chris Mooney discusses several different strategies in the political 'war' against science.[6] These are pursued against a general background of 'undermining science', as when politicians follow creationists in deriding Darwinian evolution as 'just a theory'. The success of this tactic depends on the public's ignorance of what scientific theories are and how they work. It ignores the way scientific theories interlock and, if valid, mutually support other theories: the strength of the theories of evolution and of AGW derives from their seamless fit with all other relevant theories.

There's a long history of the suppression of scientific findings when they do not support the plans of the politicians. There are 'smear' campaigns directed against individual scientists with the aim of discrediting their findings. There is the rigging of the process through packing scientific advisory boards with sympathetic members, and misrepresentation of the results, through distortion, cherry-picking, or outright falsehoods. There is a consistent reliance on the 'outliers' and their contrarian 'sound science', of the kind we saw in relation to the climate sceptics. Finally, there are the tactics of 'magnifying uncertainty' and 'merchandising doubt'.

A disturbing example of the success of such tactics is the continuing lack of regulation, especially in the United States, of a variety of health-endangering chemicals in consumer products. A salient example is the 'endocrine disruptor' bisphenol-A (or BPA), which is found in everything from plastics to canned food. Scientific studies have suggested links to breast cancer and diabetes, and to a range of attention disorders in children. BPA has recently been shown to affect various hormones, even in tiny doses, in a way that is 'epigenetic': by interfering the switching mechanisms of genes, it generates effects that extend over several generations.[7] However, following the playbook of the tobacco industry in the case of smoking, the chemical industry has used its enormous wealth to block legislation that would limit the use of endocrine disruptors for the sake of public health.

Many Republican politicians argue that even if we are producing more heat-trapping gases that are making the planet warmer, that will be just fine, since after all CO_2 is a naturally occurring gas that helps plants grow. But there is general agreement among the numerous climate models that have been run that, although crop yields might increase for a time in some parts of the world, the multitude of adverse effects globally will outweigh these localised advantages by far.

Religion in U.S. Politics

The Republican war on science is a Holy War, and this is especially true of the war on climate science. The major strategist in this campaign is Senator

James M. Inhofe, who in 2003 became Chairman of the powerful Senate Committee on Environment and Public Works. In a discussion of the notorious 'Clear Skies' legislation, Inhofe claimed that global warming is 'the second-largest hoax ever played on the American people, after the separation of church and state' (Newton-Small, 2005). In his mind, the two great hoaxes are closely related, insofar as the climate sciences ignore the role played by God in controlling the climate. And yet publicly Inhofe is concerned to present himself as being well informed about the science.

The two-hour statement he delivered on the Senate floor as the new Committee Chair bears the title, 'The Science of Climate Change' (Inhofe, 2003). He begins by acknowledging the 'profound responsibility' that comes with his position as Chair, 'because the decisions of the committee have wide-reaching impacts, influencing the health and security of every American'. In retrospect he was being overly modest here, insofar as the decisions of his committee are now influencing the health and security of most people on the planet.

Proceeding from his principle that federal agencies should 'use the best, *non-political* science to drive decision-making' (emphasis added), Inhofe goes on to discuss at great length what he calls 'an important body of scientific research that refutes the anthropogenic theory of catastrophic global warming ... [and which] offers compelling proof that human activities have little impact on climate'. The ensuing discussion of 'the Kyoto Treaty', the IPCC Assessment Reports, the temperature record, and so forth shows that Inhofe has done a vast amount of reading – but it is all very selective, and the 'important body' of research turns out to be merely the work of the 2.5% of climate scientists who are sceptics, and of Seitz, Singer, the Marshall Institute, and the like. The narrative is utterly fantastic, constructed with numerous straws grasped from studies taken out of context, and darkly Orwellian in the way it exalts the contrarian fringe as 'objective' and 'sound' science while dismissing as 'phony science' everything that the climate science community has established through decades of peer-reviewed research. Thus, the first resounding conclusion of the two-hour speech is:

> The claim that global warming is caused by man-made emissions is simply untrue and not based on sound science.

Although Inhofe had selected his sources carefully, two legitimate climate scientists, Tom Wigley and Stephen Schneider, publicly protested that Inhofe had 'misrepresented' their work and their views on climate change. His response was simply to ignore the protests and to continue to cite their work without further comment – a standard tactic of the climate deniers when accused of misrepresentation.[8]

Inhofe also follows the contrarians in the argument that, even if humans are heating up the planet, more CO_2 will, in any case, be a good thing, since crop productivity is '30 PERCENT HIGHER IN A CO2-ENHANCED WORLD' (shouted block capitals in original). Which leads to the second resounding conclusion:

> C02 does not cause catastrophic disasters – actually it would be beneficial to our environment and our economy.

In accounting for why the climate science community should be saying the opposite, Inhofe again follows the lead of the contrarian fringe. He advises his colleagues in the Senate to reject any restrictive environmental legislation,

> since it is designed to satisfy the ever-growing demand of environmental groups for money and power and other extremists who simply do not like capitalism, free-markets, and freedom.... Let me be very clear: alarmists are attempting to enact an agenda of energy suppression that is inconsistent with American values of freedom, prosperity, and environmental progress.

The fundamental cultural value expressed here – the notion that capitalism, free markets, prosperity and progress must never be constrained by regulation designed to achieve social justice and protect the resource base on which all human institutions depend – helps to explain how an influential U.S. Senator can talk for two hours on 'The Science of Climate Change' and get everything so utterly wrong. But it also becomes clear that the main problem with Inhofe's 'Statement' is that scientific evidence means nothing to him, since he already knows that human activities cannot possibly have any impact on climate. They cannot because they are nothing by comparison with God's dominion over the world.

Now, Inhofe could not say that explicitly on the Senate Floor, nor does he in his recent book, *The Greatest Hoax: How the Global Warming Conspiracy Threatens Your Future*. (The home page of the publisher's website does, however, display a banner praising the site for its 'Christian stand', and carries the headline, 'Israelis: Obama Birth Certificate Is Phony'.[9]) However, when a radio interviewer asked about the relevance of his citing *Genesis* 8:22 ('as long as the earth remains there will be seed time and harvest, cold and heat, winter and summer, day and night'), Inhofe replied:

> My point is, God's still up there. The arrogance of people to think that we, human beings, would be able to change what He is doing in the climate is to me outrageous.[10]

If you believe, as many Republican politicians do, that human activities cannot possibly affect divine control of the climate, then, of course, none of the scientific evidence of anthropogenic forcing can even make sense, let alone be valid. This is a crucial issue, not just because Inhofe is still a member of the Committee on the Environment but also because so many Republicans – and especially those affiliated with the Tea Party – follow the line that the findings of the climate sciences *must* be wrong, or irrelevant, because they are incompatible with what they understand as God's dominion over the earth.

Representative John Shimkus, Chair of the House Subcommittee on Environment and Economy, follows Inhofe in advocating more carbon dioxide because 'it's plant food', and in his belief that God controls the fate of the planet. In a subcommittee hearing on adaptation policies for dealing with climate change, Shimkus quoted passages from *Genesis* and *Matthew* – 'the Word of God' which is 'infallible, unchanging, perfect' – and drew this conclusion:

> The earth will end only when God declares it's time to be over. Man will not destroy the earth; this earth will not be destroyed by a flood [from rising sea levels caused by climate change].

He then turned, though only briefly, to the science, saying:

> Today we have about 388 parts per million [of carbon dioxide] in the atmosphere: I think in the age of the dinosaurs, when we had the most flora and fauna, we were probably around 4000 parts per million. ... There is a theological debate that this is a carbon-starved planet, *not* too much carbon. (Shimkus, 2009)

Since the dinosaurs lasted some 160 million years, it is hard to tell during what period Mr Shimkus thinks that carbon dioxide concentrations were that high, but scientific estimates of CO_2 during the mid-Cretaceous period put them at between 800 and 1600 ppm (Crowley & North, 1991). His audience may have been pleased to hear that theologians have moved on from debating how many angels can stand on the head of a pin, but there is actually very little theological debate about optimal carbon concentrations for the earth's atmosphere, since most theologians are sensible enough to leave this one to the climate scientists.

House Representative Joe Barton, former chair of the House Energy and Commerce Committee (who has received over $1.5 million in contributions from fossil fuel concerns), is another powerful politician whose views on climate change are grounded in the power of God.[11] In an interview in 2009, he claimed:

> A lot of the CO_2 that is created in the United States is naturally created. You can't regulate God. Not even the Democratic majority in the US Congress can regulate God.[12]

Then there is Ralph Hall, chair of the House Committee on Science. When asked his opinion, in an interview, on anthropogenic global warming, he replied:

> I can't say it [human activity] doesn't have a percentage of effects on it … but I don't think it's the cause. I don't think we can control what God controls. We put $32 billion into it and don't see very much change.[13]

There it is again: the belief that mere humans by definition cannot have any significant effect on divine creation. The interviewer reminded him of the survey published by the National Academies of Science which showed that over 97% of climate scientists agree that human activities are leading to global warming.

> They each get $5,000 for every report like that they give out. That's just my guess. I don't have any proof of that. But I don't believe 'em. I still want to listen to 'em and believe what I believe I ought to believe.

Because these are facts about science that the chair of the Science Committee does not want to hear, he simply does not believe them. He negates or denies the integrity of the study by inventing some tale of corruption for which he does not 'have any proof', and simply believes what he believes he ought to believe, while at the same time claiming (for the sake of appearances?) to want to listen to what the scientists are saying. But even if he did listen, it would not shake his faith – and especially his firm conviction that we have no control over the climate because God controls it. Is he 'on the same page' then, the interviewer asks, as (Texas) Governor Perry, who has a theory that climate scientists are in a conspiracy to doctor the data to get funding for their research?

> I'm pretty close. I think we ought to have an honest ear to science. They can come before my committee. I always put someone to come and testify when they're testifying against it to give them the other side. I think we oughta listen to 'em. *I just don't think we oughta mind 'em.* (Emphasis added)

This is the 'fair and balanced' approach: bring in some kind of scientist and pay him or her handsomely to present the contrarian, climate sceptical side. It is only for show anyway, since while the chair may listen to the scientists who testify before his committee, he will not pay attention to a thing they say (mind 'em), because his mind is already made up. He believes what he believes he ought to believe.

The problem here stems from a refusal (made openly by James Inhofe) to acknowledge the separation of church and state by preventing one's personal religious views from impinging on one's decisions about science and public policy. Representative Paul Broun – a member of the House

Science, Space, and Technology Committee – has claimed that, whereas 'God's Word is true', evolution, embryology and the Big Bang theory are 'lies straight from the pit of Hell'. Though he said this at a private function, he makes it clear that in his case the private thoroughly determines his work as a politician:

> The Bible … teaches us how to run all our public policy and everything in society. And that's the reason, as your congressman, I hold the Holy Bible as being the major directions to me of how I vote in Washington, D.C., and I'll continue to do that.[14]

This man disqualifies himself, as do his colleagues whom we have heard say that God takes precedence over human science and shared understanding, from being a credible member of any science committee. A government that works for the good of the people has to fire such characters for ignorance and incompetence, and replace them with people who understand natural science and acknowledge that religion has its place – but not one that allows it to preempt the evidence of science when it comes to a phenomenon such as global warming.

SYNDROMES AND SOLUTIONS

The global warming deniers, those of great faith, know that God, not human beings, controls the climate. They also trust, presumably, that although the world is filled with violence and war, much of it originating in the United States, God will not resolve it by bringing extreme weather to bear, as he did with the Flood.

> And God said unto Noah, The end of all flesh is come before me; for the earth is filled with violence through them; and, behold, I will destroy them with the earth. (*Genesis* 6:13)

The more agnostic deniers and sceptics, who acknowledge that burning fossil fuels may cause problems but insist that we cannot do anything that would impede economic growth, believe in a secular form of salvation – through cleverer technologies. There are ways, so they say, for us to continue with business as usual while reducing our emissions of greenhouse gases. To the problem of meeting our energy needs they propose a variety of technological solutions: nuclear power, carbon sequestration, fracking for natural gas and geo-engineering. (There are also biofuels, such as ethanol, but in many cases these consume as much energy to produce as they ultimately deliver, and where arable land is converted from needed food crops to biofuel production, the glaring social injustice renders this option unconscionable.) What is common to these 'solutions', all of which have considerable drawbacks and attendant dangers, is a refusal to accept any natural limits on human desires, and an insistence on developing ever more complex and invasive technologies to ensure a more comfortable life (at least for those in the developed world).

Carbon Sequestration, Etc.

Carbon sequestration is an attempt to continue and extend the process of fossil fuel burning that began in earnest with the Industrial Revolution. That Revolution depended on breaking through natural limitations on the amount of energy available for human use. In the beginning, those limits were set, in the case of energy from combustion, by the amount of fuel (in the form of naturally occurring biomass) in the neighbourhood. With the advent of agriculture came the possibility of growing greater amounts of biomass than the land produced naturally. Later on coal was used as a fuel, in China long before Europe, but mining was from surface deposits, and so limited in scale. This situation lasted many centuries, until shortly before the Industrial Revolution, when Europeans found a way to combine the art of using fire with the technique of mining.

The invention of the steam engine, most efficiently fuelled by coal, and especially the steam pump, enabled in turn the development of 'deep shaft' coal mining. This combination of technologies allowed us to delve deeper into the earth than ever before, thereby reaching back in geological time to extract fossil biomass that was deposited, much of it, during the Carboniferous period – well over 300 million years ago. Having exhausted the capacity of fuel available on the earth's surface, which incorporates energy received from the sun in recent times, depending on the life-span of the vegetation in question, we now surpass that limit by capturing energy that came from the sun hundreds of millions of years ago. Does that in itself not suggest we might throw off the energy balance of the planet? A prominent fire historian describes the consequences of this move to 'industrial fire': 'The limitations on fire reside no longer in its sources – ignition and fuel – but in the sinks such as the atmosphere that must receive combustion's unbounded by-products' (Pyne, 2001, p. xvi).

Now there is fracking for natural gas, which is thrilling the United States over the prospect of greater energy independence and lower greenhouse gas emissions, as locally fracked gas takes over from imported fossil fuels. Such enthusiasm has spread to many other countries. It is true that natural gas produces per unit of energy about half the greenhouse gas emissions of coal. But the gas extracted through fracking still belongs in that 20% of fossil fuels that is our carbon budget if we are not to endanger life on earth. And there are side-effects and by-products from the process: pollution of the water, air and soil – not to mention the enormous quantities of water that the process requires, and the attendant discharges of methane, a far more effective greenhouse gas than carbon dioxide.[15]

But it is perfectly safe! spokespersons of the industry assure us. Of course, don't they all say that? – the lead paint manufacturers, the tobacco companies, the synthetic chemicals industry, the GMO producers, the nanotechnology outfits – It is not *our* product that is responsible for this damage to your health or your genes. But now that we know about the correlations between smoking and lung cancer, perhaps we can discern the pattern and appreciate the dangers of fracking sooner rather than later.

Although nuclear power has so far proved much safer than fossil fuels (less destructive of human life and the natural environment per unit of energy yielded), it nevertheless bears a distinctive aura of death. This derives in part from its origins in the development of death-dealing nuclear weapons – the atomic bombs on Hiroshima and Nagasaki – but also because exposure to lower than immediately lethal doses of radiation can lead to the development of cancers and leukaemia decades later. Cancer is in one sense a refusal of certain cells, when they are damaged, to self-destruct for the greater good of the organism – a micro-scale instance of refusal to accept natural limits.[16] Think what we are doing when we split atoms: with nuclear fission we penetrate the innermost kernel of the elements, and mimic on the surface of the earth processes that naturally take place only in its deepest core. Does our hubris metre not sound an alarm? Are we surprised when a consequence of exposure to this process is deadly because it disrupts the innermost workings of the living cell?

Then there are the by-products in the form of radioactive waste. No satisfactory and affordable solution has been found to the problem of how and where to store the growing amount of plutonium waste from nuclear power plants, which remains deadly for many thousands of years. Costly security measures have to be taken to guard against terrorist attacks and theft of materials usable for nuclear devices. When full safety features are included, nuclear power plants become even more expensive to build, and the costs of insuring them against accidents are so prohibitive that governments have to subsidise them, which they are increasingly unwilling to do.

Here is a case where the economics of the process are beginning to show the way – or rather the unviability of that particular way.

Many technophiles propose geo-engineering as the solution. We can keep on burning all the fossil fuels we want, as long as we counteract the warming by finding ways to remove or capture the excess carbon dioxide (of which bio-energy combined with carbon capture and storage is the most promising). Or else we devise means of reducing the natural warming effect of solar radiation: if the sun is too hot because we are generating so much of our own heat, we will simply 'turn it down' by sowing the stratosphere with sulphur aerosols, or some similarly untested and unpredictable method. Ever since the Romans invented and deployed the siege engine (*ingenium* in Latin – whence our word 'ingenuity'), engines have been the driving force behind our conquest of the earth. The Industrial Revolution depended on the steam engine, and the internal combustion engine has revolutionised agriculture and transportation. We have been engineering enormous projects for a long time now – so why not the whole earth and its climate?

The Spirit of Prometheus

The current momentum of economic growth and technological progress in the developed world, and in the United States especially, would have us continue with 'business as usual' until we really know that we have a problem that will cost us economically. There is no thought that such transgression of the natural limits imposed on life by the resources of a finite planet may lead to trouble. If we step back and try to get a sense of what underlies this drive to push through the limits set by the natural world, it is impossible to ignore a mythical figure from the ancient Greek tradition: namely, Prometheus. Several key features in the story of Prometheus fit our current situation so perfectly, and especially our mania for consuming energy by way of the products of modern technology, that we do well to consider them – especially in the light of the punishment that follows.

The most fruitful way to understand a myth like that of Prometheus is indicated by the Neoplatonic philosopher Sallustius (fourth century) when he writes of one ancient myth, but referring to them all: '*These things never happened, but always are*' (Sallustius, n.d.). The events of the Prometheus story never happened, but are always going on behind what people do. If we look at the stories from this perspective, we can see patterns of behaviour that we are enacting without realising their consequences.

The myth of Prometheus, whose name means 'Forethought', is complex, and comes down to us through several sources, the earliest of which is Hesiod (7th century BCE), with his *Theogony* and *Works and Days*. Prometheus, as a second-generation Titan, possesses enormous strength derived from the powers of the heavens, oceans and earth. Hesiod introduces him as being 'shifty, quick-scheming', and as one of several Titans who rebelled against Zeus and were punished for it. After Prometheus plays a trick on him, Zeus withholds fire from humans, whom Prometheus champions, in retaliation. But for Zeus, it is also a matter of justice: he denies human beings fire, and other ways of making life easier, so that they will continue to work hard in order to make a living, which is their role in the appointed scheme of things.

But the defiant Prometheus then enrages Zeus further by stealing fire and giving it to human beings, in return for which Zeus metes out two punishments. He punishes humans by having Hephaestus fashion 'a beautiful maiden' (named Pandora in *Works and Days*), who 'wrought baneful evils for human beings' by taking the lid off her famous jar, and by arranging the world so that humans have to work in order to survive.[17] (The punishment of Prometheus himself we will consider shortly.) If it seems unjust that we humans should be punished for Prometheus's crime, the reason becomes clear when we consider that Hesiod follows the story of Pandora by an account of the five successive human races – degenerating from gold through silver and bronze and warring demigods to the present race of iron. By the time we decline to the race of iron, all sense of justice has been lost and the strong get what they want by being arrogant and violent – *hubris*.[18] And hubris is, of course, the prime offense of Prometheus, when he arrogantly tries to undo the just order appointed by Zeus.

In the version of the story told by the ancient tragedian Aeschylus in *Prometheus Bound*, the protagonist transgresses the natural limits set by almighty Zeus not only by giving humans fire but also by teaching them the major techniques of survival – such as house-building and woodworking, agriculture and animal husbandry, ship-building and the mining of 'bronze, iron, silver and gold', the sciences of arithmetic and astronomy as well as the art of medicine.[19] These arts and techniques (the Greek is *technē*, the root of our world 'technology') make life more secure and comfortable, protecting us to some extent from injury, illness and death. Prometheus boasts not only of saving humankind from destruction but also of preventing mortals from foreseeing their fate – death – by sowing in them 'blind hopes'.[20] And yet being unaware of one's fate as mortal may prevent one from living life to the full, and having blind hopes is a sickness if we are so bedazzled by the Promethean spirit of technology that we fail to foresee an avoidable catastrophe.

A century after Aeschylus, a version of the Prometheus story recounted by 'the wise Protagoras' in Plato's dialogue of that name suggests significant limitations to the gifts that Prometheus bestowed upon humanity. Shortly before the creation, Prometheus saw that of all animals the humans were in need of protection through artificial means. So he resorted to theft:

> He steals from Hephaestus and Athena artful wisdom along with fire ... and gives them to humans. In this way *humanity got the survival wisdom but did not get the political*, for that remained in the keeping of Zeus.[21]

Protagoras's account speaks clearly to our current predicament: inspired by the spirit of Prometheus, humanity has become highly skilled in the technical arts of survival (while forgetting that these skills are gifts and stolen goods), but lacks the political arts that would integrate technological expertise with the art of living together in communities – and especially now that it must also live in a community that is global.

A later source for the Prometheus story is Ovid's *Metamorphoses* (8 CE), the first book of which has the Titan himself create the human race by moulding it from mud or clay. On the model of Hesiod's story of the degenerating races of men, Ovid then recounts a decline through four ages, or eras. At the nadir, a race of Giants even rises up against the Gods. The degeneration goes along with a growth in human hubris occasioned by increasing proficiency in the Promethean arts of agriculture, house-building, and mining.[22] The arrogance of humans so angers Zeus that he sends a flood to annihilate all but a single couple, Deucalion (the son of Prometheus!) and his wife Pyrrha, who ride out the flood in a boat. Deucalion then recreates men from rocks and stones, while his wife does the same for women.[23] This division of labour suggests that men are more likely than women to be driven by Promethean urges, and that the raw materials for this new human race will make it harder and less flexible than the initial generation fashioned from mud or clay.

Ovid deplores the mining that was developed during the Age of Iron because of the way its products lead to further decadence: gold is used for ostentatious luxury and iron for weapons to fight wars (mainly to get one's hands on other people's gold). The great Roman historian of natural science Pliny the Elder, who was born just a few years after Ovid's death, condemns the practice of mining for its sacrilegious aspects as well.

> We trace out all the veins of the earth, and yet, living upon it, undermined as it is beneath our feet, are astonished that it should occasionally cleave asunder or tremble:

> as though in truth these signs could be any other than expressions of the indignation felt by our sacred parent! We penetrate into her entrails, and seek for treasures in the abodes of the spirits of the dead, as though each spot we tread upon were not sufficiently bounteous and fertile for us![24]

But from the perspective of today's geosciences, we know that earthquakes are not caused by mining – or do we? It appears that, at least in earthquake-prone areas, geothermal projects that pump water deep into the earth's crust can cause tremors of over 3 on the Richter scale.[25] Fracking seems to have a similar effect. What is more, geophysicists have shown how continued extraction and combustion of fossil fuels, together with the consequent global warming, promise to occasion greater and more frequent convulsions of the earth.[26] Whatever the nature of the causal circumstances, Pliny's idea that delving into the bowels of the earth is an affront to the earth goddess resonates in many of the world's ancient cultures, and is a call whose meaning we would do well to try to fathom today.

Let us not forget that Prometheus, whose sacrilegious theft set in train all this ingenuity, was severely punished for his crime. According to Hesiod, our oldest source:

> With painful fetters Zeus bound shifty-planning Prometheus, with distressful bonds, driving them through the middle of a pillar; and he set upon him a long-winged eagle which ate his immortal liver, but this grew again on all sides at night just as much as the long-winged bird would eat during the whole day.[27]

In antiquity the liver, as the most powerful and blood-rich organ in the body, was well understood and highly regarded as 'the seat of life', and so this punishment strikes at the core of the transgressor's existence. And as we moderns insist on transgressing natural limits through too much clever technology, we may find that the source of our very vitality is being depleted on a daily basis.

From the beginning of Aeschylus's drama, Prometheus's immobility is emphasised. Hephaestus nails him to the rock with fetters and wedges that 'leave it loose nowhere', and drives 'the obstinate jaw of the adamantine wedge right through his breast'.[28] This physical immobility, which lasts throughout the play, is mirrored by a psychological fixation, by the protagonist's pride, stubbornness and utter inflexibility. For one so obstinate in the assertion of self-will against higher powers, the punishment of complete immobilisation is fitting. At the end of *Prometheus Bound*, he remains just where he was at the start, having suffered physical torment but no change of mind or heart, intractable in his defiance of the most powerful God. In thrall to Prometheus's gifts of technology, we display a similar intransigence. In our determination to persist in warming the earth's atmosphere, despite warnings from floods, droughts, hurricanes, tornadoes and other manifestations of extreme weather, we are enacting the old drama once again – blithely oblivious to the punishment in store.

Some will say we need not take the Prometheus story so seriously, since Hesiod's account of his punishment ends with his release (albeit after 'countless years of time'): with the consent of Zeus, the heroic Heracles liberates the suffering Prometheus. However, one might well be wary of a freedom effected by Heracles: not the most intelligent of heroes, he usually employs his enormous muscle power to work *against* natural forces. (Nine of his famous Twelve Labours involve overpowering wild animals or monsters by killing or capturing them.) What corresponds nowadays to rescue by Hercules would be tricky enterprises like carbon sequestration, fracking, nuclear power and geo-engineering, which allow us to persist in our ways in the unquestioning faith that new technology will provide a fix that will keep us living comfortably. All of them 'blind hopes' bestowed upon us by Prometheus. The problem lies not so much in the Promethean arts per se, most of which could be practised sustainably, but rather in the way we use them and then get carried away by the results. Here we consistently fail to cultivate that other gift from Prometheus, namely foresight.

Renewable Energy and Sufficiency

Two far better ways are open to us, if we drop the obsession with 'meeting our energy needs': we can reduce our rate of consumption by conserving energy and using it more efficiently, and we can switch to clean, renewable energy sources.

A thousand-page report published in 2012 by the IPCC shows that renewable forms of energy (bioenergy, solar, geothermal, hydropower, ocean and wind energy) can provide 80% of the world's requirements by mid-century – but only if governments actively promote them (United Nations Environment Programme, UNEP, 2012). These figures are truly remarkable (though meaningless without that support from the world's governments), since the received wisdom has been that renewable energy is incapable, owing to its dependence on weather and other local conditions, of satisfying sustained demand. This report has not received the degree of attention it deserves, but its results have to be taken seriously because of the genuinely international and non-partisan make-up of the enormous research team that conducted the study.[29]

If the shift to clean and renewable sources of energy is accompanied by increases in energy efficiency and reductions in wasteful use, an even larger percentage of 'the world's requirements' could be provided by such sources well before the year 2050, while at the same time ensuring that considerations of social and global justice are taken into account and acted on.

Environmental economists have been saying for years that the first step in the transition is for governments – first and foremost the United States – to end subsidies for the fossil fuel concerns and put the funds into research and development of clean, non-carbon, renewable energy technologies. With additional funding to enable economies of scale that would make these technologies affordable globally, the transition could be swift enough to avert disaster – not to mention a less painful expense than continuing inaction would incur. Something like this was accomplished from the late 1950s through to the 1980s when the U.S. government invested (initially for reasons of military strategy) in the microchip technology that eventually enabled the World Wide Web. When the vast infusion of government funds had made it attractive for venture capitalists to invest, the economies of scale kicked in, start-ups proliferated, and a multitude of high technology industries took off.[30]

In 2001, George W. Bush reneged on the American commitment to the Kyoto Protocol because reducing carbon emissions would 'hurt the US economy', and this canard has become a mantra for the Republicans ever since. The Kyoto Protocol had its flaws, but the Republicans have the economics all wrong. Commentators have been warning for years that, on the contrary, the United States is missing a great opportunity to be the economic leader in renewable energy technology.[31] And, of course, they were right: China now dominates the world market in solar panels, for example. Taken as a whole, Europe is the biggest investor in renewable energy, with the United States pathetically taking third place to China.

China is actually the third crucial obstacle in the way of resolving the problem of global warming. Once the United States, as the country that has emitted the greatest amount of greenhouse gases, sees its way to taking remedial action, the next most urgent task will be to ensure that China, and then India, follows suit. The problem is that in spite of its burgeoning numbers of billionaires, China is still a developing country with some 200 million living in poverty, and hundreds of millions more without adequate housing and sanitation. The country has the resources to alleviate these conditions – but only by burning its enormous reserves of coal. If the Chinese do that, the planet will burn. (But this is the topic of an article to follow this one.)

China can never be persuaded to leave its coal in the ground unless the most highly developed countries reduce their level of consumption. This is not the unmitigated evil that people fear. The change will require some psychological adjustment, and a subsequent change in our economic system. There is no reason to suppose that these measures will diminish what is really important: not short-term pleasure or happiness, but human flourishing, lives that are meaningful and fulfilled. A survey of the ancient wisdom of the world's philosophical and religious traditions reveals that not one of them claims that a fulfilled human life is to be attained by high levels of consumption of energy or goods. Indeed, they all say quite the opposite.

The almost universal lesson concerns how much is enough, and teaches that enough is less than we think. So, given a sufficiency, and perhaps a modest amount more to moderate the austerity, when faced with an enticing product of super-capitalism there are usually *nine good reasons not to buy it*:

1. You save money, in the amount of the purchase price.
2. You don't have to clean it, keep it up or repair it if it breaks.
3. No need to protect against theft, insure it or in any way secure it.
4. You never have to bother with either storing or moving it.
5. If you don't own it, you can't lose it, or waste time looking for it.
6. There is no need to dispose of it, or take the trouble to recycle it.

And when the Grim Reaper arrives to tell you that it is time to go, you realise:

7. You can't take it with you – so you certainly do not need to pack it.
8. And since you can't bequeath it if you are not the owner, your heirs can't possibly squabble over it.
9. If *nobody* ever buys it, they will eventually have to stop making it – and that will save natural resources, reduce pollution and make everybody happier (once they get used to not needing it).

It goes without saying that if people heed these reasons and gradually stop buying things they do not really need, this will weaken the hold of global capitalism and reduce exploitation of human labour and the earth.[32]

Since this is unlikely to persuade committed consumers to change their ways, it should be noted that considerations of global justice and fairness require that people in the developed world scale back their consumption to allow the rest of the world to attain a decent standard of living. Mass media now broadcast the so-called joys of advanced consumer culture, via satellite dishes, to almost all parts of the globe, so that people in the developing countries have come to think that is what they want, too. To say to them 'We're sorry, but we got here first and it's too late for you to join the party' is unlikely to convince.

Religion and Money Out of Politics

One of the first thinkers to have dealt with the question of the proper places of science and religion in society was John Tyndall, who was not only a great teacher of science but also something of a philosopher, having studied at the University of Marburg and in Berlin. One of his major contributions in this area was to restrict the traditional involvement of religious considerations in the practice of the natural sciences. In 1874, he was elected president of the British Science Association, and he delivered for his inauguration in Belfast a highly controversial presidential address – controversial because of his outspoken insistence that religion no longer be allowed to trespass on the domain of science.[33] Loud though he spoke out, Tyndall's approach was nuanced and philosophically sophisticated. He warns against deriding the religious sentiment (as the fundamentalist atheists of today like to do with such gleeful vehemence) because it enjoys what he calls 'an immovable basis ... in the nature of man'. The most pressing problem of the present, he says (and I would say, of our present too), is how 'to yield this sentiment reasonable satisfaction'. He continues:

> Grotesque in relation to scientific culture as many of the religions of the world have been and are – dangerous, nay destructive, to the dearest privileges of freemen as some of them undoubtedly have been, and would, if they could, be again – it will be wise to recognize them as the forms of a force, mischievous, if permitted to intrude on the region of *knowledge*, over which it holds no command, but capable of being guided to noble issues in the region of *emotion*, which is its proper and elevated sphere.[34]

There is a proper place, then, for the formidable force of religious feeling, but when it infringes on scientific culture it causes mischief and mayhem: not only burnings of books regarded as heretical but also burnings of their authors at the stake – witness the sad case of Giordano Bruno (of whose 'pantheism' Tyndall speaks with great admiration in his Belfast address). Tyndall goes on to stake out an inviolable place for science in society:

> All religious theories, schemes and systems, which embrace notions of cosmogony, or which otherwise reach into the domain of science, must, *in so far as they do this*, submit to the control of science, and relinquish all thought of controlling it. Acting otherwise proved disastrous in the past, and it is simply fatuous today.

What was fatuous in Tyndall's day was the rejection of Darwin's theory of evolution by believers in the literal truth of the creation stories in the Book of Genesis. The prevalence in our day of 'scientific creationism', or (another fine oxymoron) 'creation science', vividly displays the dangers of democracy when a large enough proportion of the electorate is ignorant. A Gallup Poll published in 2012 showed that 46% of Americans subscribe to a creationist view of human origins, while the figure among declared Republican voters is 58% (Gallup & Newport, 2012).

Not long after Tyndall gave his address in Belfast, a more trenchant critic of Christianity, Friedrich Nietzsche, similarly emphasized the need to keep religion from encroaching on science. Like Tyndall, Nietzsche acknowledged its prevalence in human societies and the importance of some of its functions, though he thought the control should be exercised by philosophy rather than by natural science:

> It always costs dearly and terribly when religions reign *not* as means of cultivation and education in the hands of the philosopher, but reign for themselves and as *sovereign*, when they want to be ultimate ends and not mere means among other means.[35]

There is nothing wrong, for Nietzsche, with religions, as long as they remain under the rule of philosophy – so that their energies can be directed towards saving the earth rather than destroying it. Let us consider how this might work, in practice, since religious beliefs often strongly condition people's relations with the natural world, and consequently how they treat the earth and the other forms of life with which we share it.

Since philosophy is a different enterprise from religion, it is in a position to survey the world's religions disinterestedly with a view to highlighting those aspects of them that would conduce to saving the earth rather than destroying it. Both Christianity and Islam have been criticized by environmental thinkers for denigrating the natural world and exalting its Creator, and for fostering human arrogance by encouraging dominance over the rest of creation. However, many of the relevant passages in the scriptures can be interpreted less anthropocentrically, as Muslim and Christian theologians have done in recent decades by emphasizing humans' responsibility for taking the best possible care of God's creation.

One can shift the focus to those passages in the scriptures that praise the goodness and beauty of the earth as God's creation. If the God of the three great monotheisms consistently finds his creation 'good' and 'very good', then who are we to denigrate and desecrate it?[36] This constitutes a powerful justification of the inherent worth of natural phenomena, which would give us reason not to extinguish other species with such nonchalance, as opposed to their merely extrinsic usefulness to humans, which we take as license to exploit them for our own ends. So if (as we read in *Ephesians*) 'in the dispensation of the fullness of times God might gather together in one all things in Christ, both which are in heaven, and which are on earth, even in him', we had better 'consider the lilies of the field, how they grow', and make sure that not one of them has been gratuitously destroyed by human toil.[37]

The lilies of the field, along with many other kinds of plant, have been devastated by the impact of human agriculture, which increases naturally as world population grows. Back in the 1960s, scientists like Paul Ehrlich drew attention to the dangers of burgeoning population, and especially for the natural environment. Ehrlich (with John Holdren) formulated the IPAT equation ($I = P \times A \times T$), which says that the environmental impact of human activity is a function of population, affluence and technology, remains helpful to this day. But although discussions of environmental problems used to emphasise, for good reason, the importance of restraining population growth, the issue of population steadily became less prominent – perhaps in response to disapproval from religious leaders (and especially the Pope).[38]

This is another case where one needs to question the contemporary appropriateness of passages from Holy Scripture. The Old Testament injunctions to 'Be fruitful and multiply' were perfectly apposite when the tribes of Israel numbered so few, and were in danger of annihilation. But the equating of piety with fecundity that many religions promote now deserves to be questioned. William LaFleur has called this phenomenon 'fecundism', defined as 'the attribution of religious value and significance to reproductivity, especially if large numbers of progeny were a major part of the intentionality of sex'.[39] In a world of 7 billion people, projected to rise to 9 billion over the next 30 years, the promotion of fecundism is inappropriate and irresponsible. When you look into the reasons behind enthusiasm for burgeoning populations, they are rarely reassuring: it is usually a matter of people's wanting to have many more of 'us' so that there will be relatively fewer of 'them'.

As for money and politics, it is best, as Plato urged, to insulate the rulers as far as possible from any personal profit or gain. You clothe and feed and house them, decently but not luxuriously. Just as the best doctor is an expert who practises for the good of the patient and not for his own advantage, the best rulers work to benefit the ruled and not themselves – or themselves only as members of the society as a whole. If Plato's ideal is considered impossible to realise, we can at least aim for it by getting money out of politics as much as possible. Otherwise you have a travesty of democracy, as in the United States nowadays, where in effect a plutocracy of the 1% exploits the rest of the population for their own personal gain.

A number of steps can be taken to mitigate the dangers of climate change, of which the last two are crucial.

As far as science is concerned, the 'disagreements' with climate science expressed by the sceptics and deniers have to be ignored – unless they change their ways and engage the scientific community by producing peer-reviewed research of their own. The political 'war against science' has to stop, so that scientists can pursue their research free from government intervention or censorship; and scientists have to disclose any financial interests (such as large salaries from conservative think tanks) that might bias their findings. Climate scientists need to buttress or, in some cases, restore public confidence in the validity of their work by articulating the problems we face in language comprehensible by non-experts.

In economics, the falsehood of the claim that it will cost too much to act to prevent disastrous climate change now, rather than wait until it gets really bad, needs to be exposed as such; and the consensus among environmental economists that it is better economically to take immediate action has to be communicated by the media. The system needs to become more realistic: the costs to the environment and human health of emissions from fossil fuel burning, currently treated as 'externalities' in defiance of the 'polluter pays' principle, have to be internalised, whether by a tax on carbon or some fairer and more efficient method devised by economists. Advocates of the needed shift to renewable and sustainable sources of energy have to understand the psychology of acceptance and denial on the part of the public, and formulate their messages accordingly.

People in the developed nations have to cut back on energy use and consumption of consumer goods, which are already causing misery in the developing world and stand to leave for future generations an impossibly compromised planet. If we give it a try, we may find that, contrary to our expectations, we end up happier and more fulfilled as a result – quite apart from considerations of social and international and transgenerational justice.

The world is waiting for the United States to take action to mitigate global warming – but that is where we find the two most crucial obstructions to progress: profits directing politics and religion dictating policies. The world's religions are in general well disposed towards saving the earth, as evidenced by such movements as ecotheology and other religious engagements with raising ecological consciousness, but a type of Christianity in the United States is perverting democratic processes there towards a kind of fundamentalist theocracy. A first step would be to remove the religious zealots from all congressional and state committees that have to do with making policy on the basis of the natural sciences.

Money needs to be kept out of politics as much as possible, through campaign finance reforms that make it more likely that politicians will act for the public good rather than personal profit. The task of reducing the political power of the rich fossil fuel concerns requires a top-down political solution. Grass roots movements to curb consumerism and encourage people to buy only things they really need – preferably of good quality so they will last – are admirable and can affect some degree of change in our economic system. But while some people have been able to get themselves 'off the grid' by switching to photovoltaic and solar panels and other renewable energy sources, for most of us in the developed world, it is hard to get by without electricity, and minimal consumption of gas and oil.

We need strong political leadership on this front, and voters who make it clear the direction they want to go. We need politicians who will sit down with the CEOs of the fossil fuel industries and explain to them why they cannot burn more than 20% of their enormous reserves without destroying the planet. Since the industries can hardly argue that their profits should outweigh the livelihoods of millions in the developing world and coming generations in the developed world, a constructive dialogue about how to proceed ought to be possible.

One bright ray of hope in this dismal situation comes from the Fossil Fuel Divestment campaign initiated by Bill McKibben's 350.org, which aims to persuade universities with endowments to withdraw their investments in fossil fuel industries. Their model is the successful campaign in the 1980s that persuaded universities with large endowments to withdraw their investments in stocks of companies operating in South Africa under apartheid. Some universities with smaller endowments are divesting, but the largest endowment in the world is standing firm. Although students at Harvard recently voted overwhelmingly in favour of divesting, the University's response was curt, without explanation: 'Harvard is not considering divesting from companies related to fossil fuels' (Gillis, 2012). And why not? Does one of the world's top universities have no sense of civic responsibility, no other concern than for its own wealth, at the expense of the global public good?

The World Bank recently commissioned a study from the Potsdam Institute, which was published in November 2012 under the title *Turn*

Down the Heat: Why a 4°C Warmer World Must Be Avoided (The World Bank, 2012). At 64 pages of main text, it is admirably comprehensive and concise, and should be read by anyone interested in the fate of the planet. Calculating future climate sensitivity – the ways climate will respond to the various forcings to which human activity is subjecting it – is necessarily an uncertain exercise, but the authors of this study estimate the risks in an eminently sober manner. Compare the risks of a 2°C rise, which member states of the United Nations agreed posed the maximum permissible risk, with some projections of our 'no need to worry' and 'business as usual' scenario, which estimate a rise of 5°C or even 6°C by the end of the century. The prospects of a 4°C rise are dismal enough – and if we reach one or more climate tipping points on the way, we might as well give up all hope of a habitable planet.

For the ancient Greeks, philosophy begins with *wonder*: wonder at the ways the world works and even at its very existence. The present essay in philosophy ends with wonder: wonder at the colossal stupidity of what we as a species are doing to the earth, on which we depend for our very existence.

NOTES

1. Plato (1991), *Republic*, book 1, 346c–347d; book 3, 416d–417a.
2. For an illuminating discussion of the international issues involved here, see Vanderheiden (2008, 2009).
3. IEA analysis of fossil-fuel subsidies (n.d.).
4. For an excellent overview of the political activities of Charles and David Koch, see Mayer (2010).
5. Greenpeace (n.d., p. 6.) For more information on the activities of Koch Industries, see the web page by the Center for Media & Democracy (n.d.).
6. Mooney (2006, 84ff). Mooney admits that non-Republican politicians are also guilty of abuses, though lesser and not as systematic.
7. See, for example, Diamanti-Kandarakis et al. (2009); Vandenberg et al. (2012); and Wolstenholme et al. (2012).
8. See Mooney (2006, pp. 99–100).
9. Inhofe (2012). The website: WND – A Free Press for a Free People. (n.d.)
10. Senator James Inhofe, radio interview, 'Crosstalk', Voice of Christian Youth America, March 8, 2012; audible at: Tashman (2012).
11. On Barton's income from fossil fuel concerns, see Kreighbaum (2010).
12. Quoted from Frick (2009).
13. Ralph Hall, quoted in Davenport (2011). The text of the full interview with Hall, together with further remarks by Hall, is available at Mervis (2011).
14. Representative Paul Broun, speech at a sportsman's banquet on 27 September 2012, as reported by Pearce (2012).

15. See also the series by Ian Urbina for the *New York Times*, 'Drilling Down', Urbina (n.d.).

16. On this topic, see my essay (Parkes, 2012) 'Nuclear Power after Fukushima 2011: Promethean and Buddhist Perspectives'.

17. Hesiod. (n.d.-a, n.d.-b), *Theogony*, lines 517–616; *Works and Days*, lines 47–105.

18. Hesiod (n.d.-b), *Works and Days*, lines 174–201.

19. Aeschylus (1990), *Prometheus Bound*, lines 450–505.

20. Aeschylus (1990), *Prometheus Bound*, 250-53. There is a reference to Prometheus's depriving men of knowledge of their death in Plato, *Gorgias*, 523d–e.

21. Plato, *Protagoras* 321c–d; see Lampert (2010).

22. Hesiod, *Works and Days*, lines 109–201; Ovid (1995), Book I, 6–8.

23. Ovid (1995, I, 5–6, 17–19).

24. Pliny, Club, and Holland (1847), *Natural History*, Book xxxiii, Chapter 1.

25. A Swiss (and distinctly Promethean) geothermal energy project, the Enhanced Geothermal System (EGS), was halted in Basel in 2009 after studies projected millions of dollars of damage would be caused every year by the consequent earth tremors. Not so surprising a projection since the project was sited in an earthquake zone: the city of Basel was destroyed in 1356 by an earthquake estimated to have been 6.5 on the Richter scale, Glanz (2009).

26. See, for example, McGuire (2012).

27. Hesiod, *Theogony*, lines 521–25. The image is most powerfully rendered by Rubens (1612) in his magnificent *Prometheus Bound*.

28. Aeschylus, *Prometheus Bound*, lines 52–65.

29. The study was conducted by 'an author team of 122 Lead Authors (33 from developing countries, 4 from economies-in-transition countries, and 85 from industrialized countries), 25 review editors and 132 contributing authors'. The IPCC review procedure requires that drafts produced by the authors were subject to two reviews, in the course of which '24,766 comments from more than 350 expert reviewers and governments and international organizations were processed' (ix).

30. See, for example, Shellenberger and Nordhaus (2007, pp. 122–23).

31. See Friedman (2008); Friedman and Mandelbaum (2012).

32. The insidiousness of consumerism is compounded when, as during the past few decades, it is based on incurring high levels of debt. See John Barry (2012, pp. 255ff). This excellent study, as its subtitle suggests (*Human Flourishing in a Climate-Changed, Carbon-Constrained World*), helpfully reminds us of a consideration generally ignored in the debates over what to do about global warming: namely human flourishing.

33. In those days, it was known as The British Association. The full text can be found at Tyndall (1874).

34. Tyndall, Belfast Address, pp. 60–61.

35. Nietzsche (2013), *Beyond Good and Evil*, aphorism 62.

36. *Genesis* 1: 10-31. Also: 'All things were created by God, and for him: And he is before all things, and by him all things consist'. (*Colossians* 1:16–17)

37. *Ephesians* 1:10; *Matthew* 6:28.

38. See the 1995 encyclical letter *Evangelium Vitae* ('The Gospel of Life') by Pope John Paul II (1995), and William LaFleur's (1996) response: 'Ending Fecundism:

An Open Letter to Pope John Paul II Concerning Contraception, Religion, and Ecology' (1996).

39. The term 'fecundism' was coined by Bill LaFleur; see LaFleur (1992, p. 88).

REFERENCES

Aeschylus. (1991). *Aeschylus II: Prometheus bound*. Chicago, IL: University Of Chicago Press.

American Bible Society. (1980). *Holy Bible: King James version*. American Bible Society.

Barry, J. (2012). *The politics of actually existing unsustainability: Human flourishing in a climate-changed, carbon constrained world*. New York, NY: Oxford University Press.

Center for Media and Democracy. (n.d.). Koch industries. *SourceWatch*. Retrieved from http://sourcewatch.org/index.php?title=Koch_Industries. Accessed on May 2, 2013.

Crowley, T. J., & North, G. R. (1991). *Paleoclimatology*. New York, NY: Oxford University Press.

Davenport, C. (2011). Heads in the sand. *NationalJournal.com*. Retrieved from http://www.nationaljournal.com/magazine/heads-in-the-sand-20111201. Accessed on April 22, 2013.

Diamanti-Kandarakis, E., et al. (2009). Endocrine-disrupting chemicals: An endocrine society scientific statement. *Endocrine Reviews, 30*(4), 293–342. doi:10.1210/er.2009-0002

Frick, A. (2009). Barton: We shouldn't regulate CO_2 because "it's in your Coca-Cola' and "you can't regulate God'. *Think Progress.org*. Retrieved from http://thinkprogress.org/politics/2009/05/19/41233/barton-carbon-god/?mobile=nc. Accessed on April 22, 2013.

Friedman, T. L. (2008). *Hot, flat, and crowded: Why we need a green revolution – And how it can renew America*. New York, NY: Farrar, Straus and Giroux.

Friedman, T. L., & Mandelbaum, M. (2012). *That used to be us: How America fell behind in the world it invented and how we can come back*. New York, NY: Picador.

Gallup & Newport, F. (2012). In U.S., 46% hold creationist view of human origins. *Gallup.com*. Retrieved from http://www.gallup.com/poll/155003/hold-creationist-view-human-origins.aspx. Accessed on April 23, 2013.

Gillis, J. (2012). To stop climate change, college students take aim at the endowment portfolio. *The New York Times*. Retrieved from http://www.nytimes.com/2012/12/05/business/energy-environment/to-fight-climate-change-college-students-take-aim-at-the-endowment-portfolio.html

Glanz, J. (2009). Quake threat leads Swiss to close geothermal project. *The New York Times*. Retrieved from http://www.nytimes.com/2009/12/11/science/earth/11basel.html

Global Commission on Elections, Democracy & Security. (2012). Deepening democracy: A strategy for improving the integrity of elections worldwide. *Global-commission.org*. Retrieved from http://www.global-commission.org/report. Accessed on April 22, 2013.

Greenpeace. (n.d.). Koch industries: Secretly funding the climate denial machine. *Greenpeace*. Retrieved from http://www.greenpeace.org/usa/en/media-center/reports/koch-industries-secretly-fund/. Accessed on April 22, 2013.

Hesiod. (n.d.-a). *Theogony* (H. G. Evelyn-White, Trans.), *Sacred Texts.com*. Retrieved from http://www.sacred-texts.com/cla/hesiod/theogony.htm. Accessed on April 22, 2013.

Hesiod. (n.d.-b). *Works and days* (H. G. Evelyn-White, Trans.), *Sacred Texts.com*. Retrieved from http://www.sacred-texts.com/cla/hesiod/works.htm. Accessed on April 22, 2013.

IEA analysis of fossil-fuel subsidies. (n.d.). Retrieved from http://www.iea.org/media/weowebsite/energysubsidies/ff_subsidies_slides.pdf

IEA and OECD. (2011). OECD and IEA recommend reforming fossil-fuel subsidies to improve the economy and the environment. *OECD.org*. Retrieved from http://www.oecd.org/newsroom/oecdandiearecommendreformingfossil-fuelsubsidiestoimprovetheeconomyandtheenvironment.htm. Accessed on May 2, 2013.

Inhofe, S. J. (2003). *The science of climate change: Senate floor statement*. U.S. Senate Committee on Environment and Public Works. Retrieved from http://www.epw.senate.gov/speechitem.cfm?party=rep&id=230594. Accessed on April 23, 2013.

Inhofe, S. J. (2012). *The greatest hoax: How the global warming conspiracy threatens your future*. Washington, DC: WND Books.

Kreighbaum, A. (2010). Rep. Joe Barton Likes BP – And the company likes him back with cash. *Open Secrets blog: Investigating money in politics*. Retrieved from http://www.opensecrets.org/news/2010/06/barton-likes-bp-and-they-like-him-b-1.html

Lafleur, W. R. (1992). *Liquid life: Abortion and Buddhism in Japan*. Princeton, NJ: Princeton University Press.

LaFleur, W. R. (1996). Dear Pope John Paul. *tricycle.com*. Retrieved from http://www.tricycle.com/feature/dear-pope-john-paul. Accessed on April 23, 2013.

Lampert, L. (2010). *How philosophy became Socratic: A study of Plato's "Protagoras," "Charmides," and "Republic."* Chicago, IL: University Of Chicago Press.

Macalister, T. (2012). Budget 2012: Oil and gas industry gets £3bn tax break to encourage drilling. *The Guardian*. Retrieved from http://www.guardian.co.uk/uk/2012/mar/21/budget-2012-oil-industry-tax

Mayer, J. (2010). Covert operations. *The New Yorker*. Retrieved from http://www.newyorker.com/reporting/2010/08/30/100830fa_fact_mayer

McGuire, B. (2012). *Waking the giant: How a changing climate triggers earthquakes, tsunamis, and volcanoes*. New York, NY: Oxford University Press.

Mervis, J. (2011). Ralph Hall speaks out on climate change. *Science Insider*. Retrieved from http://news.sciencemag.org/scienceinsider/2011/12/ralph-hall-speaks-out-on-climate.html. Accessed on April 22, 2013.

Mooney, C. (2006). *The Republican war on science*. New York, NY: Basic Books.

Newton-Small, J. (2005). Utilities fall short on air rules. *Tulsa World*. Retrieved from http://www.tulsaworld.com/article.aspx/Utilities_fall_short_on_air_rules/050122_Bu_E1_Utili44244. Accessed on April 22, 2013.

Nietzsche, F. (2002). *Beyond good and evil*. Cambridge: Cambridge University Press.

Ovid. (1995). *The metamorphoses of Ovid* (A. Mandelbaum, Trans.). San Diego, CA: Harcourt Brace.

Parkes, G. (2012). Nuclear power after Fukushima 2011: Buddhist and promethean perspectives. *Buddhist-Christian Studies*, *32*(1), 89–108.

Pearce, M. (2012). U.S. Rep. Paul Broun: Evolution a lie "from the pit of hell." *Los Angeles Times*. Retrieved from http://articles.latimes.com/2012/oct/07/nation/la-na-nn-paul-broun-evolution-hell-20121007

Plato. (1991). *The republic of Plato* (A. Bloom, Trans.). New York, NY: Basic Books.

Pliny, the Elder, Wernerian Club, and Holland, Philemon. (1847). *Pliny's natural history*. [London] Printed for the Club by G. Barclay. Retrieved from http://archive.org/details/plinysnaturalhis00plinrich

Pope John Paul II. (1995, March 25). Evangelium vitae. *Vatican.va*. Retrieved from http://www.vatican.va/holy_father/john_paul_ii/encyclicals/documents/hf_jp-ii_enc_25031995_evangelium-vitae_en.html. Accessed on April 23, 2013.

Pyne, S. (2001). *Fire: A brief history*. Seattle, WA: University of Washington Press.

Rubens, P. (1612). Prometheus bound. Retrieved from http://upload.wikimedia.org/wikipedia/commons/d/d6/Peter_Paul_Rubens_032.jpg. Accessed on April 22, 2013.

Sallustius. (n.d.). On the Gods and the world, chapter IV, "That the species of myth are five, with examples of each" G. Murray (Trans.), *The encyclopedia of the Goddess Athena*. Retrieved from http://www.goddess-athena.org/Encyclopedia/Friends/Sallustius/index.htm. Accessed on April 22, 2013.

Shellenberger, M., & Nordhaus, T. (2007). *Break through: From the death of environmentalism to the politics of possibility*. New York, NY: Houghton Mifflin Company.

Shimkus, J. (2009). Rep. John Shimkus: God decides when the "earth will end." *YouTube*. Retrieved from http://www.youtube.com/watch?v=_7h08RDYA5E. Accessed on April 22, 2013.

Tashman, B. (2012). James Inhofe says the Bible refutes climate change. *Right Wing Watch*. Retrieved from http://www.rightwingwatch.org/content/james-inhofe-says-bible-refutes-climate-change. Accessed on April 22, 2013.

The World Bank. (2012). Climate change report warns of dramatically warmer world this century. *The World Bank*. Retrieved from http://climatechange.worldbank.org/content/climate-change-report-warns-dramatically-warmer-world-century. Accessed on May 2, 2013.

Tyndall. (1874). Address delivered before the British Association assembled at Belfast, with additions. *Victorianweb.org*. Retrieved from http://www.victorianweb.org/science/science_texts/belfast.html. Accessed on April 23, 2013.

United Nations Environment Programme, World Meteorological Organization, Intergovernmental Panel on Climate Change, & Potsdam-Institut für Klimafolgenforschung. (2012). *Renewable energy sources and climate change mitigation: Special report of the Intergovernmental Panel on Climate Change*. New York, NY: Cambridge University Press.

Urbina, I. (n.d.). Drilling down series – Natural gas. *New York Times*. Retrieved from http://www.nytimes.com/interactive/us/DRILLING_DOWN_SERIES.html?_r=0. Accessed on April 22, 2013.

Vandenberg, L. N. et al. (2012). Hormones and endocrine-disrupting chemicals: Low-dose effects and nonmonotonic dose responses. *Endocrine Reviews, 33*(3), 378–455. doi:10.1210/er.2011-1050

Vanderheiden, S. (Ed.). (2008). *Political theory and global climate change*. Cambridge, MA: The MIT Press.

Vanderheiden, S. (2009). *Atmospheric justice: A political theory of climate change*. New York, NY: Oxford University Press.

WND – A free press for a free people. (n.d.). *WND.com*. Retrieved from http://www.wnd.com/. Accessed on April 22, 2013.

Wolstenholme, J. T., et al. (2012). Gestational exposure to Bisphenol A produces transgenerational changes in behaviors and gene expression. *Endocrinology, 153*(8), 3828–3838. doi:10.1210/en.2012-1195

A CALL FOR A NEW POLITICAL ECONOMY OF ATTENTION: MINDFULNESS AS A NEW COMMONS

Peter Doran

ABSTRACT

Purpose – *This chapter posits that we underestimate the way in which our immersion in the 'social logic' of capitalist consumption constrains our attempts to understand and respond to the ecological crises at both a personal and political level – and that both dimensions of our response are bound together.*

Methodology/approach – *Survey of literature on psychology, well-being and mindfulness.*

Findings – *How has the culture of capitalism – its psychic investment in colonizing our attention – compromised our ability to respond meaningfully to the challenges of sustainable development? In an acknowledgement of a certain closure around such themes within Western thought, I look to a point of exteriority in Peter Hershock's work, drawing on China's Chan Buddhist philosophy, for intimations of a worldview*

Environmental Philosophy: The Art of Life in a World of Limits
Advances in Sustainability and Environmental Justice, Volume 13, 111–135

ISSN: 2051-5030/doi:10.1108/S2051-5030(2013)0000013009

that challenges the West's over-commitment to forms of 'control' in favour of a cultivation of mindful and careful awareness – and an offering of unconditional attention.

Social implications – *Draws attention to a new phase of 'enclosure' in the cultural processes of capitalism.*

Originality/value of paper – *Original introduction of a critical approach to mindfulness in the debate on well-being.*

Keywords: Attention; mindfulness; political economy; capitalism; consumerism; sustainable development

Our corporate culture has effectively severed us from human imagination. Our electronic devices intrude deeper and deeper into spaces that were once reserved for solitude, reflection and privacy. Our airwaves are filled with the tawdry and the absurd. Our systems of education and communication scorn the disciplines that allow us to see. We celebrate prosaic vocational skills and the ridiculous requirements of standardized tests. We have tossed those who think, including many teachers of the humanities, into a wilderness where they cannot find employment, remuneration or a voice. We follow the blind over the cliff. We make war on ourselves. (Chris Hedges, 2012, Nation of Change, Tuesday, 10 July; http://www.nationofchange.org/how-think-1341922195)

INTRODUCTION

How has the culture of capitalism – its psychic investment in colonizing our attention – compromised our collective ability to respond meaningfully to the challenges of sustainable development? There is an emerging consensus that consumerist excess and the media complex of consumerism are bad for us, environmentally, socially and psychologically. This chapter will present an argument that we may have underestimated the way in which our immersion in the 'social logic' of capitalist consumption actively constrains our attempts to understand and respond to the ecological crises at both a personal and political level – and that both dimensions of our response are bound together. To make the case, I look to Peter Hershock's work, drawing on China's Chan Buddhist philosophy for intimations of a worldview that challenges the West's over-commitment to forms of 'control' in favour of a need for the cultivation of mindful and careful awareness – and an offering of unconditional attention.

In the stunning American movie, *Detachment* (2011), substitute teacher, Henry Barthes, drifts from school to school, classroom to classroom. During a one-month assignment in a failing public school Barthes finds a connection to the students and teachers who are all, in their own ways, experiencing a deep loss of connection, and negotiating a world so bereft of love and attention that they have become – in a way – invisible at work and at home. Screenwriter and former public school teacher, Carl Lund, has Barthes deliver a speech in his class room during a pivotal moment in the film:

> **Henry Barthes:** How are you to imagine anything if the images are always provided for you? Doublethink. To deliberately believe in lies while knowing they're false. Examples of this in everyday life: Oh, I need to be pretty to be happy. I need surgery to be pretty. I – I need to be thin. Famous. Fashionable. Our young men today are being told that women are whores. Bitches. Things to be screwed. Beaten. Shit on. And shamed.
>
> This is a marketing holocaust. Twenty four hours a day, for the rest of our lives, the powers at be are hard at work dumbing us to death. So, to defend ourselves, and fight against assimilating this dullness into our thought processes, we must learn to read. To stimulate our own imagination. To cultivate our own consciousness. Our own belief systems. We all need these skills to defend, to preserve, our own minds. (*Detachment*, 2011)

The nature of modern mass mediatized capitalism poses a direct threat to well-being at the most subtle levels of human experience: attention and intention. In so doing, a fertile ground is being prepared for a popular denial of the world's ecological predicament, and innate faculties for personal and collective resilience are being undermined. In the words of Peter Hershock (2006), through the consumption of mass media (as well as other commodities), attention is exported out of our immediate situation:

> This compromises relational depth and quality, effectively eroding presently obtaining patterns of mutual support and contribution, and triggers further and still more extensive commodity consumption. As this recursive process intensified beyond the point at which all major subsistence needs have been commodified, consciousness itself is effectively colonized. The relational capabilities of both persons and communities atrophy, situational diversity is converted into circumstantial variety, and the very resources needed to meaningfully respond to and resolve our suffering or troubles are systematically depleted. (Hershock, 2006, at 26)

In his book on *Capitalist Realism*, which opens with a chapter entitled, 'It's easier to imagine the end of the world than the end of capitalism', Mark Fisher observes that over the past 30 years, capitalism has successfully installed a 'business ontology' in which it is *simply obvious* that everything in society, including healthcare and education, should be run as a business (Hershock, 2006). Similarly, Bourdieu and Wacquant have described a 'planetary vulgate' that has taken hold in all advanced societies – a vocabulary replete with references to 'globalization' and 'flexibility', 'governance and employability', 'underclass' and 'exclusion', 'new economy' and 'zero tolerance'. Multilateral environmental negotiations on the environment and climate change have their counterpart vocabularies informed by the neoliberal paradigm, including, inter alia: 'consumption and production', 'corporate social responsibility', 'flexible mechanisms', 'economies in transition' and 'green growth'. For Bourdieu and Wacquant, the diffusion of this new planetary vulgate – from which the terms 'capitalism', 'class', 'exploitation', 'domination' and 'inequality' are conspicuous by their absence, is the result of a new type of imperialism whose effects are all the more powerful and pernicious in that it is promoted not only by the partisans of the neoliberal revolution who, under cover of 'modernization', have tried to remake the world by sweeping away the social and economic conquests of a century of social struggles, but also by cultural producers (researchers, writers and artists) and left-wing activists who still regard themselves as progressives. Comparing it to ethnic or gender domination, the authors believe that cultural imperialism is a form of 'symbolic violence' that relies on a relationship of constrained communication to extort submission (Bourdieu & Wacquant, 2001).

The upshot has been a series of highly constrained debates at the UN-sponsored negotiations that have unfolded since the first Rio 'Earth Summit' (UNCED) in 1992 that take the form of a disavowal, when it comes to acknowledging that our well-being is now staked on our ability to address the all-consuming technology of capitalism and its active colonization of a new frontier that reaches into our experience of self and other, mind and nature. As Žižek has described,[1] capitalism relies on a structural disavowal based on an overvaluing of individual belief – in the sense of inner subjective attitude – at the expense of the beliefs we exhibit and externalize in our behaviour. So long as we believe (in our hearts) that capitalism is bad, we are free to continue to participate in capitalist exchange – settling for an ironic distance. This corporate sponsored rupture, summed up in Saul Alinsky's observation that 'most people are eagerly groping for some medium, some means by which they can bridge the gap between their morals and their practices' (1969), goes to the heart of the debate on well-being, redefining prosperity and sustainable consumption. For the choices that confront us are not merely about our relations with the world and others. The choices must also embrace a much older conversation about our relations with the self.

UNDERSTANDING OUR DILEMMA: CAPITALISM'S PSYCHIC INVESTMENT

In this chapter I will explore the genealogy of the psychic investment of 'capitalism' through the process of capitalization (Nitzan & Bichler, 2009), which I describe as a 'technology of micropractices'. These practices are most visible in the outworkings of the operation of mass media, advertising and the culture of consumerism and represent the culmination of a deeply ambivalent tradition in Western thought that has resulted in a profound 'breach of faith toward everything that is' (Boss, 1965).[2] For Apffel Marglin and Bush it is this breach, first articulated by René Descartes, that not only enabled unprecedented levels of human control and manipulation of the social and natural world but also lies today behind a deep alienation and meaninglessness. Since the 16th century, 'control' has been a key strategic value informing the explosion of technological development associated with the rise of the European West and its influence over other parts of the world. Hershock notes that what we refer to generically as 'technology' is actually a particular family or lineage of technologies that has arisen and been sustained through a complex of political, social, economic and cultural forces focused on the value of exerting control over our circumstances to enhance *felt independence*. This strategic value has delivered military and ecological destruction on a scale hitherto never attained, having co-evolved with and serviced the rise of the modern nation State (Scott, 1998). Writing from a Buddhist perspective, Hershock notes that although we remain related to others and to our environment, the prevalence of control fosters a dichotomous perspective on that relationship – a splitting into the objective and subjective – that facilitates treating our relations with others as either actually or potentially instrumental:

> No longer intimately continuous with all things – that is, related internally – gaps open in what I can attend or hold in careful awareness. By ignoring what intimately connects who 'I am' with what 'I am not', I render myself liable to being blindsided – subject to accidental or fateful events of the sort that cause the experience of trouble or suffering. Asserting independence through exercising technologically mediated control almost paradoxically renders us subject to new vulnerabilities.[3]

Hershock believes that the single most important long-term cost of convenience and control – in the context of the availability of global commodities – is an overall erosion of relational quality resulting in a mounting incapacity for appreciation and contribution. He observes an overwhelming trend in contemporary commodity consumption practices in the compression of the temporal and spatial scope of consumption that it mimics a digital transition from *not having* to *having* – a transition that disallows any complex, improvised relational ground being navigated or any qualitative shift in relationality being initiated and sustained. Noting considerable implications for the nature and dynamics of the public sphere, he describes the market valorization of convenience and choice as signalling both a general narrowing of our horizons of personal responsibility and, over time, a severe compromise of relational capability and attunement. Each act of commodity consumption marks a smooth and efficient paving over of opportunities for developing the complex attentive and relational skills associated with contributory virtuosity. And, in the process of handing ourselves over to the purveyors of expertly designed and manufactured goods, services, knowledge products and meaning, we are rendering ourselves increasingly in need of expert, globally mediated, care. Degraded environments, then, are inseparable from degraded consciousness, in a dual pattern of degradation that at once devalues what is experienced and lowers experiential quality. Hershock goes further and suggests that the colonization of consciousness is in many ways a more critical threat to our possibilities for realizing truly liberating environments than is the depletion of soil, the fouling of our rivers, lakes, seas and skies. The mass media have become the primary system through which the attention economy manages to be a net producer of dramatic entropy or situations in which no matter what choices we make, they will not ultimately make much of a difference. This has among its salient effects the conservation of an uncontested space for the perseverance of liberal individualism and an ontological bias towards existents rather than relationships. To counteract our dramatic impoverishment and attentive atrophy, he commends the direct cultivation of appreciative and contributory virtuosity. Hershock looks to the cultivation or rehabilitation of a form of freedom understood as virtuosic skill in improvising meaningful interdependence. Desire, then, need not be solely viewed as a source of frustration arising from self-centred attachment or craving. Rather, with cultivation, desire can become a crucial factor in the immediate realization of an unprecedented responsiveness (Hershock, 1996).[4]

At the heart of the dilemma outlined here is a decisive break in our understanding or appreciation of *the ethical* (under the sign of modernity), a break that has serviced the growing gap between our collective ability to *discuss* the *ethical* and *devise* effective strategies. These envisaged strategies would contribute to the cultivation of embodied micropractices consistent with the translation of our recognition of 'ecological boundaries' into demands for consistent lifestyle choices and practices for the great transition (The New Economics Foundation, 2009).

Foucault believed that a moral code per se is inadequate. For an action to be 'moral' it must not be reducible to an act or a series of acts conforming to a rule, a law or a value. There is no moral conduct, for Foucault, that does not also call for the forming of oneself as an ethical subject; and no forming of the ethical subject without 'modes of subjectification' and an 'ascetics' or 'practices of the self' that support them (Bennett, 2001; Foucault, 1995). In the course of his work, Foucault makes clear that the modalities of self-formation are always potentially prescribed. In his work on *Discipline and Punish: The Birth of the Prison* (Foucault, 1995), he discusses 'biopower' as the normalizing power of disciplinary practices that write the law into the body. However, in the later work, Foucault holds open the prospect for self-formation that cannot be reduced to the possibilities of normalizing power such as that exercised by the institutions of capital. He insists on the possibility of reflective modification of the sensibility of the self by the self, positing that there is no self without discipline, no discipline that does not also harbour opportunities for the arts of the self and no effective ethics without such an aesthetic project. For Foucault a moment of freedom survives within subjectivity after all, at least when it is not reduced to a purely intellectual formation. Foucault resignifies freedom by locating it in relation to a historically situated rationality and a recalcitrant body. He tries to find a way to speak of the moments of individual self-direction that persist inside disciplinary power and to do so outside of a Kantian vocabulary. To engage 'the limits that are imposed on us' is, says Foucault,[5] at the same time to 'experiment with the possibility of going beyond them' (Foucault, 1991).[6]

For Bennett, sensibility or refinement of new assemblage of sensible primorida, while culturally encoded and temperamentally delimited, is still educable to some degree. Like the code dimension of ethics, techniques of sensibility-formation are concerned with governing and refining behaviour. The difference is that these techniques respond to subtle norms of admirable behaviour and thought; they address the question of which modes of perception and which styles of comportment, and not simply which actions, are most laudable. She posits that this refinement might even make for a more resilient and careful approach to ethics (Foucault).[7]

CAPITAL AS POWER AND MASS CONSUMERISM

In an essay calling upon artists to pursue the truths of the times we live in through honest, socio-politically responsive work, Scottish playwright David Greig argues that one of the key roles of theatre in our times is to resist 'the management of the imagination by power'. Here, Greig paints a picture of the influence of capital and power on the core cultural mythology:

> The institutions of global capital manage the imagination in the first instance through media institutions. Hollywood cinema, the television and newspapers of the great media empires like Fox and CNN. These forms create the narrative superstructure around which our imagination grows. In this way we learn to think along certain paths, to believe certain truths, all of which tend, in the end to further the aims of capital and the continuance of economic growth. Once the superstructure is in place, our own individual creativity will tend to grow around it and assume its shape so that the stories we tell ourselves, the photographs we take and so forth, are put in the service of the same narratives and assumptions. ... Very few imaginations are totally colonized, just as very few are totally liberated. In most minds there is a constant back and forth – a dialogue between challenge and assumption like waves washing against a shoreline. ... By intervening in the realm of the imaginary, power continually shapes our understanding of reality. (Greig, 2007; Nada-Rajah, 2010)

For an understanding of the colonizing power of capital in the realm of culture we can do no better than look to the contemporary theoretical work of Jonathan Nitzan and Shimson Bichler for whom capitalization is 'the central institution and key logic of the capitalist *nomos*'.[8] It is the algorithm that generates and organizes prices. Specifically, they understand capitalization as the representation of the present value of a future stream of earnings: it tells us how much a capitalist would be prepared to pay *now* to receive a flow of money *later*. A concept that was perfected in the 20th century – notably with the rise of the corporation – its principle characteristic is universality. In other words, the calculation of value based on a claim on future earnings can be applied to everything. Nitzan and Shimson explain:

> Nowadays, every expected income stream is a fair candidate for capitalization. And since income streams are generated by social entities, processes, organizations and institutions, we end up with the 'capitalization of everything'. Capitalists routinely discount human life, including its genetic code and social habits; they discount organized institutions from education and entertainment to religion and the law; they discount voluntary social networks; they discount urban violence, civil war and international conflict; they even discount the environmental future of humanity. Nothing seems to escape the piercing eye of capitalization: if it generates earning expectations it must have a price, and the algorithm that gives future earnings a price is capitalization.[9]

The all pervasive influence of capitalization suggests to Nitzan and Bichler that capitalism seems able to shape 'preferences' as effectively as any authoritarian regime and is able to mould habits and instil fears. Indeed, capitalism does more by virtue of its ability to make these 'preferences' and outcomes sufficiently predictable for capitalists to translate them into expected profit discountable to present value. On the face of it, they add, liberal capitalism is all about 'individuality' and 'free choice'. And yet, the so-called individual consumer ends up being part of a collectively managed mob. And here's the rub: the Friedmanite[10] individual may feel 'free to choose' her location in the distribution, but the distribution itself is shaped by the power institutions and organizations of capitalism. And it is this shaping – that is the very creation of a predictable 'representative' consumer – that gets capitalized.[11] The consumer has every choice under the sky except the choice of whether to consume! The global expenditure on advertising in 2008 reached nearly half a trillion US dollars, equivalent to more than $80 dollars each for every man, woman and child on the planet.[12]

One of the contributions of the Nitzan and Bichler's thesis on 'capital as power' is their demonstration of how our understanding of the operation of capital must incorporate any power arrangement, institution and process that systematically affects the flow and temporal pattern of earnings, because this is how the capitalist views the world. In other words, it is impossible to distinguish contributions to accumulation as clearly 'economic'. Instead we must include institutions ranging from politics and culture to the family. It follows that capitalists exert control within relevant communities in order to maintain the status quo and protect the value of the assets they own or seek to provoke changes in order to generate and divert earnings (Cochrane, 2008, p. 114). Moreover, building on the work of Veblen, Nitzan and Bichler show that – for the capitalist – there is no absolute benchmark against which accumulation may be judged as successful or unsuccessful. Rather, the success of accumulation is a matter of *differential comparison*, that is capitalists and their enterprises seek only to 'beat the average' in a defining and unending intra-capitalist struggle over trades, takeovers and mergers, together with the pursuit and capture of assets not already capitalized. Evaluation of success occurs within markets as participants buy and sell on the expected ability of the vested interests to turn the 'social control' exercised by dominant capital in the 21st century into earnings. Markets therefore constitute empirical representations of control. Cochrane comments:

> It is these groupings within dominant capital whose actions have the greatest influence on political economic developments. Their struggles for differential accumulation are massive exercises of power that continually order and reorder society. (2008, p. 116)

DiMuzio (2007) has illustrated the extent to which dominant capitalist groups depend on broader social, cultural and political processes in their attempts to beat the average returns and redistribute a larger share of earnings away from their competitors:

> For example, the corporate earnings of Apple Computers Inc. do not simply depend upon the ability to produce its range of iPods or other goods and services by directing the labour of its workers. Rather, their corporate earnings, and the willingness of investors to bid up the price of existing shares, depend upon a whole range of factors that the corporation may wish to influence: the perception that portable mp3 players are a necessity; their ability to press for new markets and trade agreements with other legal jurisdictions; the ability of the state to punish violators who infringe their intellectual property rights; the quality of their lobbyists; their public reputation; accounting practices and standards; the ability of its consumers to access credit; the ability to influence anti-trust legislation and so on.

It is this understanding of accumulation that leads Nitzan and Bichler to argue that 'power is both the means and the end of accumulation' (Nitzan & Bichler, 2002, p. 10). Corporations exert whatever power they can muster over society, politics and culture in order to generate earnings and beat the average. DiMuzio notes that, if Nitzan and Bichler are correct, behind this movement is a whole history of corporate power deployed to 'restructure society and affect its overall development' (Nitzan & Bichler, 2002, p. 38). At the heart of their thesis is a claim that anything – including any dimension of human creativity and behaviour – can be targeted for capitalization through enclosure, expropriation and commodification. Indeed, as DiMuzio concludes, for Nitzan and Bichler, the modern corporation has become an incipient form of the State, insofar as firms control ever more facets of life and planetary resources. Power itself has become commodified.

CONSUMERISM AND THE 'POLITICAL TECHNOLOGIES OF INDIVIDUALS'

For a detailed insight into the relationship between the modern state and the consumer-citizen, we must now turn to Foucault's work on governmentality and biopolitics. Foucault's notion of governmentality points to a reciprocity in the constitution of power techniques and forms of knowledge. The semantic linkage of governing ('*gouverner*') and modes of thought ('*mentalité*') suggests that it is not possible to study the technologies of power without an analysis of the political rationality underpinning them. There are two sides to governmentality. Governments define a discursive field in which exercising power is 'rationalized'. This process – often with governments in a tacit response to direct or indirect corporate influence – involves a series of problematizing operations, wherein governments identify/define and represent 'reality' and offer solutions to a series of perceived 'problems' (problematization). A paramount consideration, of course, is the security, reproduction and continuation of the State itself through a combination of geopolitics and political economy. In advanced capitalist economies, the problematic pursuit of economic growth has come to be rationalized in a number of ways, including through a claim that prosperity and opulence facilitate and support the human need for symbolic interaction, linkage with the provision of public services such as education and health, and finally, government interest in economic and social stability.[13] The reproduction of the advanced industrialized state necessitates a reproduction of conditions that support a set of co-dependent subject–object relations, with individuals located in, moving between and latterly, conflating the roles of 'citizen' and 'consumer' (Doran, 2010).

Foucault sought to show how the modern sovereign State and modern autonomous individuals co-determine each other's emergence. For example, in his study of the Chicago School as a social form, Foucault suggested that the key element in the School's approach is their consistent expansion of the economic form to apply to the social sphere, thus eliding any difference between the economy and the social (Lemke, 1997, p. 7). A key feature of the neoliberal rationality is the congruence it endeavours to achieve between a responsible and moral individual and an economic–rational individual. Neoliberalism encourages individuals to give their lives a specific entrepreneurial form. Critically, in this transposition of individual life choices onto entrepreneurial forms, and rendering 'economic' those areas of life that were previously extra-economic, now to be decided on the basis of economic efficiency, a close link is forged between economic prosperity and 'self-care' or personal well-being (Lemke, 1997, p. 12).

Reith notes that consumers are expected, within neoliberal States, to govern themselves through their consumption habits, with the ideal of consumer sovereignty based on autonomous individuals shaping their own trajectories through their actions in the marketplace. These self-determining agents are responsible for their own welfare, security and future happiness independent of wider systems of social support (Reith, 2007, p. 33). He adds: '... the ideologies of free choice and *consumer sovereignty actually become the regulatory principles of modern life*' (emphasis added) (Reith, 2007, p. 48).

Rose's work on psychology, power and personhood provides a useful theoretical departure point for a 'genealogy of subjectification' (1998). To write such a genealogy is to seek to unpick the ways in which the self functions as a 'regulatory ideal' in many aspects of our contemporary forms of life, including 'our systems of consumption' (Miller & Rose, 2008). The 'subject of consumption' is the individual who is imagined and acted upon by the imperative to consume. Rose comments: 'A genealogy of subjectification takes [this] individualized, interiorized, totalized, and psychologised understanding of what it is to be human as the site of a historical problem, not as the basis for a historical narrative'.[14] This genealogical work follows Foucault's interest in 'our relation to ourselves' (Foucault, 1988). It is a genealogy of a 'being's' relation to itself and the technical forms that this has assumed, according to Rose:

> Our relation with ourselves, that is to say, has assumed the form it has because it has been the object of a whole variety of more or less rationalized schemes, which have sought to shape our ways of understanding and enacting our existence as human beings in the name of certain objectives.[15]

Of particular interest in this context is Foucault's keenness to direct our attention to the ways in which strategies for the conduct of conduct frequently operate through trying to shape 'self-steering mechanisms'. It is through these technologies and mechanisms that modern individuals experience, understand, judge and conduct their selves. Of particular importance in his genealogy of subjectification is the role of the so-called 'psy-' sciences, including psychology. These, according to Rose, have acquired a peculiar penetrative capacity in relation to practices for the conduct of conduct, and play a key role in our contemporary regime of subjectification and its unification under the sign of the self. Indeed, a critical history of the psy-disciplines must take as its object our contemporary regime of the self and its identity.[16] Rose suggests that we might learn most about the relation between the vicissitudes of capitalism and the rise of the psychological disciplines by examining the political, institutional and conceptual conditions that gave rise to the formulation of different notions of the economy, the market and the labouring classes. He invites us to attend to the ways in which these conditions problematized different aspects of existence, and to analyse the ways in which these problematizations produced questions to which the psychosciences could come to provide answers (legitimating their claims to authoritative knowledge production in the process).[17] There is no more pressing set of problematizations for the field of green economy and sustainable consumption than those psychological practices that enclose the citizen as sovereign consumer, constantly presented to herself as an essential node in the growth complex of capitalist production and consumption. Rose observes:

> ... and we should explore the ways in which the psychosciences, in their turn, transformed the very nature and meaning of economic life and the conceptions of economic exigencies that have been adopted in economic activity and policy.[18]

For Rose, psychology can be viewed as a form of technology, as an ensemble of arts and skills entailing the linking of thoughts, affects, forces, artifacts, techniques that do not simply manufacture and manipulate, but which, more fundamentally, *order being*, frame it, produce it, make it thinkable as a certain mode of existence that must be addressed in a particular way. He observes how, in liberal democratic societies, 'norms and conceptions of subjectivity are pluralistic, but the condition of possibility for each version of the contemporary subject is the birth of the person as a psychological self, *the opening of a space of objectivity located in an internal "moral" order, between physiology and conduct, an interior zone with its own laws and processes that is a possible domain for a positive knowledge and a rational technique* (emphasis in original)'.[19]

Branches of the psy-sciences facilitate the translation of important features of the capitalist complex, notably the objects of capitalization,[20] in a way that they 'become psychological', in that they are *problematized* or *rendered simultaneously troubling and intelligible* in terms that are infused by psychology. As we have noted,[21] the Friedmanite individual may feel 'free to choose' a location in the distribution, but the distribution itself is shaped by the power institutions and organizations of capitalism. And it is this shaping – that is the very multifaceted creation of a predictable 'representative' consumer – that gets capitalized. Luke comments that under corporate capitalism the plannable life course of all individuals *qua* consumers becomes a capital asset in that the 'consummative mobilization' of production directly boosts the productivity, profitability and power of corporate capital's increasingly automated industries (Luke, 1999).

Psychologists have come to participate in the fabrication of contemporary reality – in producing calculable transformations of the social world – providing the language to establish translatability between politicians, lawyers, managers, bureaucrats, businessmen and individuals. Rose helps to explain this process:

> Convinced that we should construe our lives in psychological terms of adjustment, fulfilment, good relationships, self-actualization, and so forth, we have tied ourselves 'voluntarily' to the knowledges that experts profess, and to their promises to assist us in the personal quests for happiness that we 'freely' undertake.[22]

Calculability, or rather the problematization of calculability of individuals, is one of the contributions of the social role of psychology: 'Truth thus takes a technical form: truth becomes effective to the extent that it is embodied in technique'.[23]

Consumption is one of the key sites for the deployment of contemporary presuppositions concerning the self. Expertise has forged alignments between broad socio-political objectives, the goals of producers and the *self-regulating propensities* of individuals. A complex economic terrain has taken shape, in which the success of an economy is seen as dependent on the ability of politicians, planners, and manufacturers and marketers to differentiate needs, to produce products aligned to them and to ensure the purchasing capacity to enable acts of consumption to occur. While political authorities can only act indirectly upon the innumerable private acts that comprise consumption, it is the expertise of market research, of promotion and communication, underpinned by the knowledge and techniques of subjectivity, that provides the relays through which the aspirations of ministers, business and the dreams of consumers achieve mutual translatability. In rendering the internality of the human being into thought, in rendering it simultaneously visible and practicable, the psycho-sciences have made it possible for us to dream that we can order our individual and collective existence according to a knowledge/technique that fuses truth and humanity, wisdom and practicality.

CAPITALIST REALISM IN THE 21ST CENTURY

In the Western experience, it seems the Kantian-inspired bar on consideration of the sensuous affect as integral to the agency side of pure morality has, it seems, contributed to a disembodied moral life that contributes to an ethos of pleasure and comfort-seeking as a poor substitute for 'the good life'. The 'goods life' has trumped the good life. Another Buddhist commentator, David Loy, describes 'attention' as the basic commodity – the fundamental target of capitalization as the production complex seeks to convince us that the solution to our *dukka*[24] or nagging sense of lack is the next purchase.

Commodity culture has turned the relation between morality and the sensuous on its head. This is most visible in the impoverished value systems of narcissism in celebrity culture and the accompanying forms of violence inflicted on the body. In an essay on Michael Jackson – 'so consumed by self-loathing he carved his African American face into an ever-changing Caucasian death mask' – Chris Hedges observes that the fantasy of celebrity culture is not designed simply to entertain (2001). It is designed, rather, to drain us emotionally, confuse us about our identity, make us blame ourselves for our predicament, condition us to chase illusions of fame and happiness, and keep us from fighting back. Rowe coined the term 'attention economy', explaining that the basic resource of the new economy is not something provided *to* the consumer but something provided by the consumer to the capitalist complex, namely 'mindshare' (Rowe, 2001). 'But what if there's only so much mind to share?' he asks (Rowe, 2001). Might the social depression and stress that accompanies the culture of consumerism be traced to the commodification of cognitive space – a new frontier in the long history of enclosure ... an enclosure of the cognitive commons, the ambient mental atmosphere of daily life.

If Nitzan and Bichler offer a compelling account of the totalizing drive of dominant 'capital as power',[25] then Fisher's *Capitalist Realism* serves as the user handbook for the consumer immersed in the universe of unfreedom posing as unlimited choice. Capitalism, writes Fisher, in a nod to Marx and Engels's *The Communist Manifesto*[26] – is what is left when beliefs have collapsed at the level of ritual or symbolic elaboration, and all that is left is the consumer-spectator, trudging through the ruins and the relics. This vision of control and communication no longer relies on subordination but extends an open invitation for all to interact and participate.

Fisher's notion of capitalist realism encompasses much more than the quasi-propagandistic way in which advertising functions. It is what he describes as a pervasive atmosphere, conditioning not only the production of culture but also the regulation of work and education, and acting as a kind of invisible barrier constraining thought and action. Here we recognize the penetrative power of capitalization – reaching across institutions where forms of social control can deliver 'attention' and recruit it in the race to 'beat the average' return.[27]

In the process capitalism engages in an 'amoral affective engineering', preferring to articulate injunctions in terms of 'health' rather than ethics. Fisher writes: 'Morality has been replaced by feeling. In the 'empire of the self' everyone 'feels the same' without every escaping a condition of solipsism'.[28] Noting the prevalence of widespread mental health problems – including those among students in UK universities – Fisher calls for a conversion of these problems into effective antagonisms, describing affective disorders as 'captured discontent' caused by Capital. He concludes:

> Furthermore, the proliferation of certain kinds of mental illness in late capitalism makes the case for a new austerity, a case that is also made by the increasing urgency of dealing with environmental disaster. Nothing contradicts capitalism's constitutive

> imperative towards growth more than the concept of rationing goods and resources. Yet it is becoming uncomfortably clear that consumer-self regulation and the market will not by themselves avert environmental catastrophe. There is a libidinal as well as a practical case, to be made for this new ascesis. If… unlimited license leads to misery and disaffection, then limitations placed on desire are likely to quicken rather than deaden, it.[29]

In his discussion on the 'social logic of consumerism' and delinking the prevailing understanding of prosperity as the accumulation of material wealth, Tim Jackson notes a consensus in the academic literature on the existence of a 'social recession' in modern western society.[30] The consensus holds that there are rising rates of anxiety and clinical depression, increased alcoholism and binge drinking, and a decline in morale at work. Berardi[31] notes that the technical definition of depression is the deactivation of desire after a panicked acceleration and calls on us to see depression not as a mere pathology, but also as a form of knowledge. Citing James Hillman, Berardi recalls that depression is a condition in which the mind faces the knowledge of impermanence and death. Suffering, imperfection, senility, decomposition: this is the truth that can be viewed from a depressive point of view. Drawing on the work of Deleuze and Guattari, Berardi asserts that when dealing with a depression the challenge is not to bring the depressed person back to normality, to reintegrate behaviour in the universal standards of normal social language. The goal, rather, is to change the focus of the sufferer's depressive attention, to re-focalize, to deterritorialize the mind and the expressive flow. The goal is to offer the possibility of seeing new landscapes, to overcome the obsessive and repetitive refrain. At the level of society, he anticipates a reconsideration of the notion of wealth and its association with purchasing power, so that a new emphasis might be placed on enjoyment. For it is in the disciplinary culture of modernity that has equated pleasure and possessing that many of our problems have their origin. And economic thinking created scarcity and privatized social need in order to make possible the process of capitalist accumulation. In the days to come, Berardi anticipates that politics and therapy will be one and the same.

CONCLUSIONS: MINDFULNESS AS COMMONS

We have known for some time that modernity and its exemplary mode of material transmission in the form of capitalism has only progressed by imposing collateral damage on society and nature. Indeed, for Carlisle,

Henderson, and Hanlon (2009), well-being *is* the collateral damage. They agree that the science of well-being and its critique are, despite their diversity, re-connected by, and subsumed within, the emerging environmental critique of modern consumer society.

Eckersley has linked static or declining levels of well-being in 'modern' societies because they focus primarily on economic and materialist concerns, to the exclusion of other values, and are characterized by rampant individualism and consumerism (1994). The renowned sociologist Zygmunt Bauman has documented the history of Western modernity as a continuous obsessive and compulsive modernization in every sphere of life, with profound consequences for how we live, act and think (Bauman, 1998a, 1998b; Bauman, 2000, 2001). His description of mobile and de-territorialized capital under the sign of 'liquid modernity' captures many of the dynamics that bring uncertainty and transcience into modern life. One of the vehicles is consumerism and its culture of disposability, wherein consumers are guided by aesthetic interests, not ethical norms. In the absence of ideals or recipes for a 'good life', the result for more and more individuals has been an experience of mental depression and feelings of impotence and inadequacy (Carlisle, 2008). For Apffel-Marglin and Bush,[32] and Hathaway and Boff (2009), the provocations forced by the global environmental crises – and their implication in a series of social pathologies – invite an investigation that must revisit the origins of a paradigm of knowledge and power codified in the 17th century, and which quickly established a homology with the expansion of the market economy and the rise of the modern State. The historical resolutions – arising from the need to restore a sense of certainty in the wake of the spiritual-cum-epistemological movements in the Renaissance, and provoked by the Reformation – left deep traces in the paradigm of modernity that was to emerge.

As Jackson and Victor (2011) have noted, capitalism – due to the 'productivity trap' (growth = jobs = social stability) – has no easy route to a steady state position. Its natural dynamics push it towards one of two states: expansion or collapse. Jackson (2011) believes that any new economy will have to take three steps: (i) establish and impose meaningful resource and environmental limits on economic activity; (ii) develop and apply a robust macro-economics for sustainability; and (iii) Redress the damaging and unsustainable social logic of consumerism. On the latter, Jackson has noted that the profit motive stimulates a continual search by producers for newer, better or cheaper products and services ('creative destruction'; Schumpeter, 1934) and the way in which the continual production of novelty is intimately linked to the symbolic or communicative role that material goods play in our lives. Noting that the social logic that locks people into materialistic consumerism as the basis for participating in the life of society is extremely powerful, Jackson adds that it is also detrimental ecologically and psychologically, contributing to a 'social recession'. He advocates structural change designed to address the social logic of consumerism, consisting of: (i) dismantling the perverse incentives for unproductive status competition; and (ii) New structures that provide capabilities for people to flourish – and particularly to participate meaningfully in the life of society – in less materialistic ways (Jackson 2011, p. 163). One avenue will be the development of non-consumerist ways of understanding and being in the world. It is envisaged that a less materialistic society will increase life satisfaction; and a more equal society will lower the importance of status or positional goods.

The 'social recession' manifests in a number of symptoms that flow from a disintegration of social ties or what Zygmunt Bauman (2002) has described as *social liquidity, including* 'consumer society' wherein all things, goods and people are treated as consumer objects. Liquid society is the result of a process that has accelerated from the early 1980s along with neoliberalism and globalization; it is a mobile, transient, precarious society in which the disintegration of social ties reaches levels that have been hitherto unknown. Bonaiuti (2012, p. 41) has linked this disintegration to: (i) the spread of individualistic behaviours and to positional competition; (ii) a contribution to the loss of well-being in contemporary societies; (iii) a loss of resilience of social organization when faced with external stress (economic or ecological); and (iv) *to a clue to comprehending why contemporary societies seem to show little reaction when confronted with the multidimensional crisis we are facing*.

Many of us are now familiar with the argument that advanced capitalism is hitting up against both planetary boundaries (Rockström et al., 2009) and 'social limits' (Raworth, 2012) associated with myopic behaviour and hyper-individualism. But what if the 'social recession' is not only undermining our psychological well-being but also undermining our ability to respond to the ecological crisis? As Bauman (2005, p. 117) suggests, 'Imagining the possibility of another way of living together is not a strong point of our world of privatised utopias'. Perhaps we need to pay more attention to the way we experience the deep socio-cultural patterning of advanced capitalism. Perhaps caring for the self is a necessary pre-requisite for and accompaniment to our collective response to the larger systemic crises.

Tracing the rise of the post-Fordist brand of newly invigorated capitalism in 1980s Britain, Rutherford (2008) describes how the new capitalism extends commodification into the realms of subjective life and invades the space of creative living (Winnicott)…'Just as early industrial capitalism enclosed the commons of land and labour, so today's post-industrial capitalism is enclosing the cultural and intellectual commons (both real and virtual), the commons of the human mind and body, and the commons of biological life'. Paul Virno (2004) has argued that the productive force of post-Fordist economic activity is 'the life of the mind'. Not just cognition but also intuition and the symbolic world of the unconscious, where communication is non-verbal. Rutherford detects a tragic dimension in the culture of capitalism that has depoliticized class while heightening the inequalities and social gulf between classes. Consumption may offer the pleasurable pursuit of desire but it is also a mass symbolic struggle for individual social recognition, which distributes shame and humiliation to those lower down the hierarchy: 'The pain of failure, of being a loser, of being invisible to those above, cuts a deep wound in the psyche' (Rutherford, 2008, p. 14). In turn, this kind of stress dramatically increases our vulnerability to disease and premature death.

In a report on Mental Health, Resilience, Inequalities (2009), the World Health Organisation described mental health as a fundamental of the resilience, health assets, capabilities and positive adaptation that enable people to cope, to flourish and to experience good health and social outcomes. It is also a key pathway through which inequality impacts on health. There is overwhelming evidence that inequality is a key cause of stress in itself and also exacerbates the stress of coping with material deprivation. It is noted that communities across Europe place a high value on well-being just as the limitations of consumerism are being more widely reflected on in relation to children, family life and the basis of civil society.

Noting considerable implications for the nature and dynamics of the public sphere where we must, finally, negotiate and engage with the crises of ecology, Hershock describes the market valorization of convenience and choice as signalling both a general narrowing of our horizons of personal responsibility and, over time, a severe compromise of relational capability and attunement. Each act of commodity consumption marks a smooth and efficient paving over of opportunities for developing the complex attentive and relational skills associated with contributory virtuosity. And, in the process of handing ourselves over to the purveyors of expertly designed and manufactured goods, services, knowledge products and meaning, we render ourselves increasingly in need of expert, globally mediated, care.

Degraded environments, then, are inseparable from degraded consciousness, in a dual pattern of degradation that at once devalues what is experienced and lowers experiential quality.

Hershock goes further, suggesting that the colonization of consciousness is in many ways a more critical threat to our possibilities for realizing truly liberating environments than is the depletion of soil, the fouling of our rivers, lakes, seas and skies. The mass media have become the primary system through which the attention economy manages to be a net producer of dramatic entropy or situations in which no matter what choices we make, they will not ultimately make much of a difference.

At least two primary sets of responses to such investigations are emerging. Both elements will have to form part of what I am calling a political economy of attention for the age of the anthropocene. At the macro-level of the economy and society, the totalizing drive of the neoliberal phase of capitalism – whose rise accompanied the decades that preceded and followed on the heels of the 'Earth Summit' in Rio in 1992 – has reached a point of exhaustion both in terms of the need to revisit the capitalist ideology of 'growth' (and its role in concealing the institutionalization of inequality across countries and within countries) and to confront the challenge of redesigning an economic system in the service of people and respectful of the planet's ecological and atmospheric boundaries. As Jackson has outlined,[33] meaningful steps to a transition to a sustainable economy must establish and impose meaningful resource and environmental limits on economic activity, develop and apply a robust macro-economics for sustainability and redress the damaging and unsustainable logic of consumerism.

Integral to the macro-response is an emerging recognition that a parallel and urgent challenge for Western citizen-consumers is the need to recast the notion of prosperity in a new language of flourishing and well-being rooted, in part, in a reclamation of 'attention' and 'somaesthetics' from the complex of capitalization. Institutional support, consistent with a new social logic, will also be required to support a new and holistic appreciation of the human being as opposed to the self-interested 'radical subjectivism'[34] cultivated by and in the service of the market. As Sachs observes in his chapter on 'The Mindful Society', an integral part of restoring balance to our engagement with society, health and the economy, will be a restored quality of mindfulness as a key element in silencing the 'relentless drumbeat of consumerism' (Sachs, 2011). Mindfulness and contemplative practices (yoga, tai chi, meditation) are already embedded in American classrooms from Princeton to Westpoint, where students begin their classes in silence. For Apffel-Marglin and Bush, the emergence of contemplative practices in our universities is an entirely appropriate response to the 21st century 'onto-epistemological situation we find ourselves in', one that requires new tools for empathy and inquiry, tools that allow us to inquire into a world with which we share our 'interbeing' and support a recovery of ethics.[35] Hershock describes mindfulness practices as an alternative technology – an alternative to our technological bias towards control and wanting. For control has silenced the things and people sharing our world, making it impossible for them to spontaneously and dramatically contribute to our narration.[36] Unfortunately, a strategic silencing of alternative ways of seeing the world and the human being has been one of the major achievements of unfettered capitalism, a strategic silencing that effectively patrols what can and cannot be contemplated in the course of current global environmental diplomacy.

Twenty years after the first Rio 'Earth Summit' (UNCED), much of the optimism generated by the political and media spectacle of high-level earth politics has dissipated. The proliferation of multilateral environmental agreements and summits that followed 1992 points to an unprecedented achievement in international diplomacy but also to the gap that continues to exist between the aspirations of the 'children of Rio' and a world haunted by an all pervasive fear that much needed change – notably at the level of society, communities and individual lifestyles in the developed world – has been too little, too late. An acceleration of global environmental diplomacy has taken place alongside, and apparently with little impact on, an unprecedented era of globalized trade, investment and the ascendancy of financialized capitalism that has left few parts of the world untouched. In 20 years of credit-fuelled spending and consumer confidence, more people, perhaps in all of human history, got to witness the sublime beauty and complexity of this lonely planet on television screens, video players and Hollywood movies. The same media complex today bears tidings of a global recession, the spectacle of stalled climate negotiations, rumours of an impending energy crisis and a widespread collapse in confidence in the political class. We are at once captivated by our dilemmas and yet condemned to an intimate distancing from our bodies, minds and the earth under the spell of capitalist realism.

NOTES

1. Cited in Fisher (2009, p. 13).
2. See Apffel-Marglin and Bush (2005).

3. See Hershock (2006, pp. 90–91).
4. See Hershock (1999, p. 79, 2006, pp. 90–91).
5. See Bennett (2001, pp. 144–146).
6. *Ibid.*, p. 146.
7. *Ibid.*, p. 150.
8. See Nitzan and Bichler (2009).
9. *Ibid.*, p. 158.
10. A reference to the leader of the Chicago School of economic thought, Nobel laureate, Milton Friedman. Friedman has been a highly influential figure for his association of capitalism with freedom. He regarded agency and freedom in the context of the market as much more than instrumental values but as ends in themselves. In libertarian terms, Friedman viewed market activities as existential goals. See Marglin (2008).
11. See Nitzan and Bichler (2009, pp. 160–161).
12. The chief aim of global advertising, described in Kanner and Gomes as 'the largest single psychological project ever undertaken by the human race', is to sell consumerism itself and shore up a consumerist, addictive culture. Kanner and Gomes describe the construction of a 'consumer self', resulting from the 'merciless distortion of authentic human needs and desires'. See Kanner and Gomes (1995).
13. See Jackson (2009). Jackson discusses the psychological downsides or pathologies associated with materialism and affluence. Collectively these unintended consequences go by the name of 'affluenza'.
14. See Rose (1998, p. 23).
15. *Ibid.*, p. 28.
16. *Ibid.*
17. *Ibid.*, p. 48.
18. *Ibid.*, p. 45.
19. *Ibid.*, p. 65.
20. See Nitzan and Bichler (2009).
21. *Ibid.*, p. 160.
22. See Rose (1998, p. 77).
23. *Ibid.*, p. 162.
24. A Buddhist term associated with suffering arising from ignorance, craving and insatiability.
25. See Nitzan and Bichler (2009).
26. In *The Communist Manifesto* of 1848, Marx and Engels described how capitalism had drowned the most heavenly ecstasies of religious fervour, of chivalrous enthusiasm in the icy water of egotistical calculation.
27. See Fisher (2009).
28. *Ibid.*, p. 74.
29. *Ibid.*, p. 80.
30. See Jackson (2010, p. 86).
31. See Berardi (2009, pp. 214–215).
32. See Apffel-Marglin and Bush (2005).
33. See Jackson (2011, p. 162).
34. See Marglin (2008, p. 64).

35. See Apffel-Marglin and Bush (2005, pp. 21–22).
36. See Hershock (1999, p. 280).

REFERENCES

Alinksky, S. (1969). *Reveille for radicals* (p. 94). New York: Vintage Books.

Apffel-Marglin, F., & Bush, M. (2005). *Healing the breach of faith toward everything that is: Integration in academia.* Retrieved from http://www.contemplativemind.org/programs/academic/summer05/Apffel-Marglin_Bush.pdf

Bauman, Z. (1998a). *Globalization: The human consequences.* Cambridge: Polity Press.

Bauman, Z. (1998b). *Work, consumerism and the new poor.* Berkshire: Open University Press.

Bauman, Z. (2000). *Liquid modernity.* Cambridge: Polity Press.

Bauman, Z. (2001). *The individualized society.* Cambridge: Polity Press.

Bauman, Z. (2002). *Society under siege.* Cambridge: Polity Press.

Bauman, Z. (2005). *Liquid life.* Chichester: Wiley Publishing.

Bennett, J. (2001). *The enchantment of modern life: Attachments, crossings, and ethics* (pp. 144–146). Princeton: Princeton University Press.

Berardi, F. B. (2009). *The soul at work: From alienation to autonomy.* Boston, MA: MIT Press.

Bonaiuti, M. (2012). Degrowth: Tools for a complex analysis of the multidimensional crisis. *Capitalism Nature Socialism, 23*(1), 30–50.

Boss, M. (1965). *A psychiatrist discovers India* (pp. 102–121). Oswald Wolf.

Bourdieu, P., & Wacquant, L. (2001). Neoliberal newspeak: Notes on the new planetary vulgate. *Radical Philosophy, 108*, 1.

Carlisle, S. (2008). *'Modernity and its consequences for wellbeing', cultural influences on health and wellbeing in Scotland.* Discussion Paper 6. Retrieved from http://www.ogilvie.fastmail.co.uk/healthyfuture/bauman.html

Carlisle, S., Henderson, G., & Hanlon, P. W. (2009). "Wellbeing": A collateral casualty of modernity? *Social Science and Medicine, 69*(10), 1556.

Cochrane, D. T. (2008). Castoriadis, Veblen, and the 'power theory of capital'. Paper presented at the 2008 Great Lakes Political Economy Conference, Ottawa, 23–24 April, pp. 114, 116.

DiMuzio, T. (2007). The "art" of colonisation: Capitalising sovereign power and the ongoing nature of primitive accumulation. *New Political Economy, 12*(4), p. 517, 522.

Doran, P. (2010). Is there a role for contemporary practices of askesis in supporting the transition to sustainable consumption? *International Journal of Green Economics, 5*(1), 22.

Eckersley, R. (1994). Dialogue on despair: Assessing the West's cultural crisis. *The Futurist, 28*(2), 16.

Fisher, M. (2009). *Capitalist Realism: Is there no alternative?.* London: Zero Books.

Foucault, M. (1988). *The care of the self – History of sexuality* (*Vol. III*). New York: Vintage Books.

Foucault, M. (1989). An ethics of pleasure. In J. Johnston & S. Lotringer (Eds.), *Foucault Live* (p. 266). Cambridge MA: Semiotext(e).

Foucault, M. (1991). What is enlightenment? In P. Rabinow (Ed.), *The Foucault reader: An introduction to Foucault's thought.* London: Penguin.

Foucault, M. (1995). *Discipline and punish: The birth of the prison.* New York: Vintage Books.

Greig, D. (2007). Rough theatre. In R. D'Monte & G. Saunders (Eds.), *Cool Britannia? British political drama in the 1990s*. Basingstoke: Palgrave Macmillan.

Hathaway, M., & Boff, L. (2009). *The tao of liberation: Exploring the ecology of transformation* (pp. 103–104). New York: Orbis Books.

Hedges, C. (2001). *The world as it is: Dispatches on the myth of human progress* (p. 40). New York: Nation Books.

Hedges, C. (2012). *Nation of change*. Retrieved from http://www.nationofchange.org/how-think-1341922195. Accessed on 10 July 2012.

Hershock, P. D. (1996). *Liberating intimacy: Enlightenment and social virtuosity in Ch'an Buddhism*. New York: State University of New York.

Hershock, P. D. (1999). *Reinventing the wheel: A Buddhist response to the: Information age* Albany: State University of New York Press.

Hershock, P. D. (2006). *Buddhism in the public sphere: Reorienting global interdependence*. London: Routledge.

Jackson, T. (2009). *Prosperity without growth? The transition to a sustainable economy*. London: Sustainable Development Commission

Jackson, T. (2010). *Prosperity without growth: Economics for a finite planet*. London: Earthscan.

Jackson, T. (2011). Societal transformations for a sustainable economy. *Natural Resources Forum*, *35*(3), 155–162.

Jackson, T., & Victor, P. (2011). Productivity and work in the new economy: Some theoretical reflections and empirical tests. *Environmental Innovation and Societal Transitions*, *1*(1), 101–108.

Kanner, A., & Gomes, M. (1995). The all-consuming self. In T. Roszak, M. E. Gomes, & A. D. Kanner (Eds.), *Ecopsychology: Restoring the earth, healing the mind* (pp. 80, 83). San Francisco: Sierra Club Books.

Lemke, T. (1997). *The birth of biopolitics – Michael Foucault's lecture at the College de France on neoliberal governmentality* (pp. 7, 12). Retrieved from http://www.thomaslemkeweb.de/engl.%20texte/The%20Birth%20of%20Biopolitics%203.pdf

Luke, T. W. (1999). *Capitalism, democracy and ecology: Departing from Marx* (p. 72). Illinois: Illinois University.

Marglin, S. (2008). *The dismal Science: How thinking like an economist undermines community*. Boston, MA: Harvard University Press.

Marx, K., & Engels, F. (1848). *The communist manifesto*. London: Merlin Press.

Miller, P., & Rose, N. (2008). *Governing the present*. Cambridge: Polity Press.

Nada-Rajah, R. (2010). *A review of 'environmental justice' research in the UK*. Retrieved from http://environmental-justice.com/research/

Nitzan, J., & Bichler, S. (2002). *The global political economy of Israel* (p. 10). London: Pluto Press.

Nitzan, J., & Bichler, S. (2009). *Capital as power: A study of order and reorder*. London: Routledge.

Raworth, K. (2012, February). *A safe and just space for humanity*. Oxfam Discussion Paper. Oxford: Oxfam International Paper. Available at http://www.oxfam.org/sites/www.oxfam.org/files/dp-a-safe-and-just-space-for-humanity-130212-en.pdf

Reith, G. (2007). Gambling and the contradictions of consumption: A genealogy of the "pathological" subject. *American Behavioural Scientist*, *51*(33), 33, 48.

Rockström, J., Steffen, W., Noone, K., Persson, Å., Chapin, F. S., III, Lambin, E., ..., Foley, J. (2009). Planetary boundaries: Exploring the safe operating space for humanity. *Ecology and Society*, *14*(2), 32. Retrieved from http://www.ecologyandsociety.org/vol14/iss2/art32/

Rose, N. (1998). *Inventing ourselves: Psychology, power and personhood* (p. 23). Cambridge: Cambridge University Press.

Roszak, T., Gomes, M. E., & Kanner, A. D. (Eds.). (1995). *Ecopsychology: Restoring the earth, healing the mind.* San Francisco, CA: Sierra Club Books.

Rowe, J. (2001). Carpe callosum. *Adbusters*, *9*(6).

Rutherford, J. (2008). The culture of capitalism. *Soundings: Journal of Culture and Politics*, *38*, 8–18.

Sachs, J. (2011). *The price of civilization: Reawakening American virtue and prosperity* (pp. 161–183). London: Random House.

Schumpeter, J. (1934). *The theory of economic development*. London: Transaction Publishers.

Scott, J. C. (1998). *Seeing like a state: How certain schemes to improve the human condition have failed.* London: Yale University Press.

The New Economics Foundation. (2009). *The great transition: A tale of how it turned out alright*. London: The New Economics Foundation.

Virno, P. (2004). A grammar of the multitude: For an analysis of contemporary forms of life. Cambridge MA: MIT/Semiotext Foreign Agent Series.

World Health Organisation/Friedli, L. (2009). *Mental health, resilience, inequalities*. WHO Europe Office.

TERRITORY IN A WORLD OF LIMITS: EXPLORING CLAIMS TO OIL AND ICE

Cara Nine

ABSTRACT

Purpose – *To determine the normative philosophical legitimacy of territorial claims to the Arctic high seas.*

Methodology/approach – *In this chapter I sketch a philosophical guideline for determining the scope of territorial rights based on established theories of territorial claims.*

Findings – *The scope of territorial rights should be limited to a geographical domain within which a group can establish a site of justice. Because currently a site of justice is not possible in the Arctic high seas, no state can extend a territorial claim to that area.*

Implications – *If adopted, this theory would prohibit the establishment of claims to the Arctic high seas made by countries such as Russia, Denmark (via Greenland) and Canada.*

Keywords: Arctic; territorial claims; resource rights; environmental justice

Environmental Philosophy: The Art of Life in a World of Limits
Advances in Sustainability and Environmental Justice, Volume 13, 137–155

ISSN: 2051-5030/doi:10.1108/S2051-5030(2013)0000013010

Although international law currently limits state oceanic claims to a 200-mile economic zone beyond their natural landed borders, parts of the Arctic are claimed by many states and indigenous groups. Countries such as Russia, Denmark (via Greenland) and Canada are working to establish a basis of their claims to the region on the basis of continental shelf features. The prize to be gained by this territorial expansion is billions of tons of oil and natural gas deposits.

While disputing parties press their own claims, the underlying issue raised is not *which* group should control the Arctic region, but rather if *any* group should. The default position regarding the Arctic has been one of peace and cooperation. Yet the division of the Arctic into territorial slices is taken as inevitable. In this chapter I sketch a guideline for determining the scope of territorial rights. The scope of territorial rights should be limited to a geographical domain within which the group can establish a site of justice. Because a site of justice is not possible currently on the ocean floor or surface of the Arctic, no group can extend a territorial claim to that area. Similar arguments can be made about the Antarctic and, I will argue, to deep underground resources.

My analysis begins by investigating the normative groundwork for a general right to territory. Chaim Gans draws a distinction between the 'right to territorial sovereignty and … how the location of territorial sovereignty is to be determined (Gans, 2001)'. A political unit (such as a nation) may have a right to territorial sovereignty based on its legitimate claim to political independence. Once the political unit has this right, another question must be addressed: *where* should they exercise territorial sovereignty? (Simmons, 2001). The latter category described by Gans describes *particular* rights to territory: a particular right to territory held by a particular group over a particular portion of the Earth. Although the question of particularity is important to territorial rights theory, in this chapter I shall focus on another category of rights: the *general* right to territory. A general right describes the claim that a group has to territory, even if we cannot locate that territory or even clearly identify a claimant.

In the first half of this chapter I present and defend the general collective right to territory from the fundamental principle that each person has an equal right to the secure access to the object of his or her basic human needs. According to this principle, a collective may acquire territorial rights from the commons if the acquisition is necessary for the provision of members' basic needs, or can be used by the collective (without spoilage) and does not prevent others from meeting their basic needs. These criteria apply to the taking of territory from the commons. The taking of territory that is already held by others (such as in the case of secession) requires further argument that I do not explore here. In the second half of the chapter, I argue that according to these criteria, the scope of territorial rights is limited to those areas where a collective can establish a *site of justice*. A site of justice is a geographical domain within which the space and resources of the domain are organized to establish just institutions. Consequently, territorial claims over Arctic and Antarctic regions and deep underground resources are not justified because these are not sites of justice.

It is important to note here that because the main focus of this chapter is to establish the scope of general territorial rights, I do not engage in the defense of these criteria against other theoretical foundations of territorial rights. Because I am working within a cosmopolitan conception of rights, I take it for granted that any argument for an exclusive claim to a good is subject to an overriding concern for the equal basic needs of all persons. In the final section a number of possible objections from the perspective of alternative theories of territorial rights are discussed.

BASIC NEEDS AND TERRITORY

In this section I explore whether collective territorial rights are required for persons to meet their basic needs. If so, then there is good reason to think that acquisition of territory from the commons is justified. In order to understand this theoretical position, we must understand territorial rights. First, a territorial right is distinct from a property right. A territorial right includes the powers to establish the rule of law within and regarding a geographical region, and this rule of law is binding on the persons within the region. A property right, by contrast, does not include this political power. The primary function of a territorial right is to establish justice. By contrast, the primary function of a property right is to aid the owner in her pursuit of her conception of the good.

Theories of pre-institutional rights, rights that a person or group may have prior to or against a political institution, describe the legitimate acquisition of property from the commons. The 'commons' describes the status of goods under circumstances where there is no government or established private property. (I further describe the commons in the second half of the chapter.) According to foundational property theorists such as Hugo Grotius, Samuel Pufendorf and John Locke, it is always permissible to take goods from the commons for private use if the taking is necessary for self-preservation. Under these theories, persons are fundamentally equal and have a duty to preserve their selves and others. This is why it is permissible to take from the commons (to take from what all others have access to) in order to preserve oneself. Theories of property also go beyond merely sustaining human life – we are allowed to take from the commons for other reasons. For example, according to Lockean theory, one may take from the commons in order to create added value or to avoid the tragedy of the commons. However, *any* privatization of the commons is subject to an important constraint: one may not take from the commons if it endangers the self-preservation of others. I am not permitted to privatize the only water hole in the desert, for example, because this privatization endangers others' lives. So in other words, one may acquire exclusive rights over goods from the commons if the acquisition:

1. is necessary for the provision of basic needs to sustain a human life, or
2. can be used by me and/or my dependents and does not prevent others from meeting their basic needs.

The above criteria are traditionally only applied to property rights. However, it can be argued that they are also relevant for territorial rights.[1] Private property rights alone are not enough to meet the needs of individuals to sustain a human life. Property rights provide an important source of control over the environment that help individuals meet their basic needs, but property rights do not exhaust the kinds of control that must be asserted over our environment in order to meet our needs. I contend that a collective has the right to acquire territorial rights from the commons if it is necessary to secure access to the objects of members' basic needs.[2]

What will help us explain and justify territorial rights from a fundamental concern for basic human needs is a fuller discussion of these needs. This discussion should acknowledge that our ability to meet our basic needs is reliant on complicated social institutions. To interject necessary complexity into the discussion of relevant human needs, we can turn capabilities theory. Made prominent in the work of Amartya Sen, capabilities theory entails the claim that primary moral value is found in the capability to do and to be what one has reason to value (Sen, 1980). Capability theory articulates, from a normative point of view, what features are necessary for a life to be *human* as opposed to merely 'alive'. On this view, what marks a human life is the ability to exercise human agency – the freedom to choose and achieve things.[3]

The achievement of minimal human agency requires that a person has access to threshold levels of freedom and well-being. In particular, human agency requires physical and psychological health; sufficient security to be able to act; a sufficient level of understanding of the options one is choosing between; autonomy; and decent social relations with at least some others (Brock, 2009, pp. 66, 67). Autonomy can be understood to include things like the capacity to form and endorse one's own ideas. But for our purposes, we should highlight elements of autonomy that Martha Nussbaum identifies as 'control over one's environment'. Control over one's environment is divided into two kinds:

> Political. Being able to participate effectively in political choices that govern one's life; having the right of political participation, protections of free speech and association.
>
> Material. Being able to hold property (both land and movable goods), not just formally but in terms of real opportunity; and having property rights on an equal basis with others; having the right to seek employment on an equal basis with others; having the freedom from unwarranted search and seizure. (Nussbaum, 2000)

Control over the environment on this view is both intrinsically and instrumentally valuable. It is intrinsically valuable because it is an essential part of the dignity of a human being. In the words of Hernado de Soto, 'The core idea is that of the human being as a dignified free being who shapes his or her own life in cooperation and reciprocity with others, rather than being passively shaped or pushed around be the world in the manner of a "flock" or "herd" animal' (Nussbaum, 2000). A *human* life is characterized by the person's capacity to control, in conjunction with others, his or her material and political environment. Without this control, a person does not have the power to shape her own life in accordance with her own desires.

Control over the environment is instrumentally valuable because of its role in securing the means to fulfil other basic needs. Political and material control are central to the individual's capacity to achieve basic functionings in other areas such as health, social relations, security and the capacity to form and endorse one's own ideas. This instrumental benefit is seen in two different kinds of control over the environment: holding property rights and political participation rights. A minimal control of property is vital to one's ability to secure one's person against incremental harm and malnutrition. Property rights are vital to getting people out of poverty, which is the main factor in cases of malnutrition and disease. For example, a right to a domicile ensures secure access to housing which provides 'shelter, dignity, and a means of accumulation'. In most cases property doesn't just secure a current income, but also provides assets and security with which a person can generate income (de Soto, 2000; Meinzen-Dick, 2009).

Beyond their intrinsic value, political participation rights also provide an instrumental benefit. Individuals need political control over their larger physical environment in order to provide for their other basic needs. Basic infrastructure, such as reliable public transportation, access to clean water and access to health care are essential collective provisions that control individuals' larger physical environment. Individual members of collectives need to be able to control, to a certain minimal extent,[4] political decisions regarding these broader physical structures, to direct collective actions so that basic services are provided. Moreover, reaping the benefits of property rights and political participation requires both to be in place. In order for property rights to fulfil their functions in the above respects, individuals must be able to participate in political decisions to ensure that their property rights remain secure. This requires that a person is a member of a political group that has political control over the geographical region where that person is domiciled.

So far, I have explained that individuals require certain basic functionings in order to live a minimally human life. Moreover, each person *equally* needs these basic human functionings – each person equally requires a life that meets this minimum threshold of functioning (although going beyond the minimum is preferred). Thus, for our purposes, a first basic principle is that situations where any persons' basic needs are not met as the result of taking from the commons are unjust.

Applying this principle to the current state of human affairs, we can derive a general justification for the taking of territory from the commons. First, the principle reveals the need for joint *collective* aims and actions. As the function of territorial rights is the establishment of the rule of law, territorial rights will normally be collectively held by a self-determining group. The taking of territory by a collective, as opposed to an individual or institution, is justified on the basis of the principle of basic needs. The current state of human affairs is one in which persons cannot achieve basic functioning on their own. Conditions of population density, resource scarcity, and the contemporary nature of economic survival make us dependent on the cooperation of others to meet our basic needs.

Over half of the world's population live in urban areas and work in industries that do not directly feed, clothe or shelter those persons. Instead, most people engage in complex market transactions to meet those needs. I rely on others to grow food and to supply it in markets in order to meet my basic nutritional requirements. My apartment was built by others, and my income (with which I provide for myself and my dependent's needs) is sourced through the government's income and property taxation scheme. In general, meeting needs in contemporary society involves complex social cooperation. In turn, contemporary social cooperation requires that individuals are able to control secure access to their environment; social cooperation relies on persons having property rights adjudicated by a set of institutions designed to coordinate intra-societal and trans-generational use of resources within the area. Through this secure system of rights individuals can form legitimate expectations allowing them to navigate the collective world in ways that will provide access to the objects of their basic needs.

Second, territorial authority over property is required to the extent that it constitutes a necessary precondition for *a stable system of property rights* for individuals. Property rights without a common adjudicator are indeterminate and prone to cause conflict. As John Locke (1632–1704) has explained in his *Second Treatise of Civil Government*, the implication is that individual property owners should submit to a common judge. While this does not preclude my neighbour and I submitting to different political institutions, it does imply that my neighbour and I can agree on a common judge for adjudicating property conflicts. According to Locke, in order for property claims to be determinate and not to give rise to conflict between owners, the owners must submit to a common judge to adjudicate their disputes over the many boundaries of property rights.[5] Persons who live in geographical proximity to one another are more likely to fall into confusion and conflict over the boundaries of their many adjacent property rights. (I am more likely to trespass on my neighbour's land than on the land of a person who is not geographically proximate to me.) Since my neighbour and I are likely to need a common judge, and my neighbour and her neighbour are going to need a common judge, and so forth, a geographically oriented adjudicator is quintessential.

Third, *encompassing* territorial coordination is necessary to provide secure access to the objects of individual basic needs. The coordination of rights over goods geographically requires institutional structures that have the authority and capacity to encompass a wide array of geographically located activities and interests into account. The physical coordination of infrastructure, including roads, maintenance of public spaces, city and regional planning together with zoning rights, will generally function best when there exists a form of encompassing institutional oversight. This oversight is more effective if it covers all of the activities performed in geographical proximity to one another. For example, institutional oversight (including the management of public safety, corporate rights, and rights to education) is necessary to direct the building of a foreign-owned chemical waste plant away from nearby housing and public schools. Smaller customary jurisdictions may not have the scope to take each of these considerations into account – they may have authority over the school but not over foreign investment activities, for example generally, a larger, encompassing statutory jurisdiction provides better coverage in these various respects.

A fourth avenue of support is that *trans-generational* territorial coordination is necessary to provide secure access to the objects of basic needs. Meeting basic needs requires collective trans-generational management of natural resources for the purpose of providing future citizens with the means to preserve themselves and to flourish – we depend upon the infrastructure created by previous generations for the provision of our current basic needs, and future generations will likewise depend upon our own collective efforts. Maintenance of urban infrastructure (maintenance of roads, means of public transportation like rails systems, waterways, sewage treatment facilities and so on) is vital for the present and future functioning of the community that uses that infrastructure to live, work and play. Note that meeting basic needs requires the cooperative trans-generational management of large sets of goods which are not only privately owned but also include communal and un-owned goods. Even in rural and less technologically developed eras, large-scale resource management (such as the building and maintenance of bridges and aqueducts) is required in order for a society to coordinate trans-generational actions necessary for the individuals to have secure access to the objects of their basic needs, especially projected forward for the provision of future generations. For example, well-coordinated ecological maintenance of previous generations allows for tailored forms of agriculture in the current generations.

Fifth, there is a need for *large-scale* collective resource management. Long-term, large-scale planning regarding resource use is central for the continued political self-determination of a society. This collective self-determination is in turn valuable for individuals' control over their environment. The political expression of a community and its members requires that their surrounding environment is amenable to their means of expression. For instance, an agricultural society will train its future members in appropriate farming techniques. The fulfilment of the society's desire to engage in their traditional economic activities requires that the environment allows for them to do so – that they have land to grow their crops. Agricultural systems imply social coordination to ensure that development of land and water resources were sustainable for future generations. Given the increasingly complex patterns of resource use developed by societies, the management of these resources for the provision and self-determination of future generations is even more apparent.

Additionally, control over the symbolic use and function of space requires collective control. The provision of public meeting spaces as well as the protection of religious sites involves the collective rule of law over those spaces. Without collective territorial authority, the collective does not have control over the development of spaces essential for their freedom of thought, association and of religion. Moreover, large-scale resource management is absolutely vital for the globally coordinated use of resources. Given global conditions of climate change and the scarcity of resources, how one society uses resources affects the ability of persons around the world to meet their basic needs. Collective agents with recognized and effective authority over resource use within their domain are better able to coordinate with each other regarding appropriate use of resources – as opposed to agents without territorial authority.

It can be concluded that that collective territorial authority is indeed essential for the ability of persons to meet their basic needs. At minimum, in order for individuals to meet their basic needs, they must form collectives that, through the political self-determination of the members, maintain institutions that provide the necessary external circumstances to this end.[6] These collectives meet members' basic needs through territorial rights (the rule of law over a geographical region) if they establish a well defined site of justice within that region.

THE SCOPE OF TERRITORIAL RIGHTS: QUESTIONING RIGHTS TO THE ARCTIC AND UNDERGROUND RESOURCE RIGHTS

The above arguments establish that a general right to territory is justified in order to provide individuals with secure access to the objects of their basic needs. With this understanding of territorial rights, we can begin to understand their scope.

It is useful to compare the circumstances under which the general right to territory is justified with the circumstances under which it is not. Natural law theorists such as John Locke, Hugo Grotius and Samuel Pufendorf used the concept of common ownership to describe a set of rights and liberties appropriate to the circumstances of simplicity and plenty. As a matter of fact, common ownership of the world is a state without property rights. It is not a state of group property rights, since all individuals did not own the world together. Instead, there were no property rights at all – nobody was excluded from anything. All individuals had a liberty to use the bounty of the earth according to their needs and desires, but they could not exclude others from doing the same. The global community in the state of common ownership can best be understood 'by analogy with the situation of invited guests at a buffet banquet … they do not each have a right, shared with others, to all the food provided. Rather the food is just there for the taking: each guest is free to take what he or she wishes, and wrongs nobody by so doing, provided what is taken has not already been claimed by someone else' (Buckle, 2002).

According to Dutch jurist and scholar Hugo Grotius (1583–1645), persons have a use-right – the right to use things for the purpose of one's preservation – in the state of common ownership. Once a thing is occupied, others do not have a liberty to claim it. This occupation only extends to 'use', such as the eating of apples or the use of a shelter while one occupies it. While the concept of 'use-rights' seems sufficient to guarantee self-preservation for persons who do not face problems of scarcity or population density, Grotius acknowledges that mere use-rights are insufficient for self-preservation once the bounty of the earth is not enough to fill this need (Grotius, 1868). In his view, there is a progression of natural rights over objects over time as persons become more dependent on secure access to those objects in order to secure self-preservation. As needs (due to population density and resource scarcity) outstrip the natural bounty of the earth, they will need to cultivate crops to provide sustenance. The successful cultivation of crops depends on the long-term control of a piece of land – such that others are prevented from using the land in a way that destroys the crops (Grotius, 1625).

So initially, agents are justified only in taking limited items from the commons. The taking from the commons is regulated by fundamental principles of natural equality and the self-preservation and benefit of humankind. In other words, one may acquire exclusive rights over goods that were formerly part of the commons, that is if the acquisition is necessary for the provision of basic needs, or can be used without spoilage and does not prevent others from meeting their basic needs.

However, this initial stage admits to only a limited set of exclusionary rights over goods. One may only acquire rights over goods that meet the above conditions. Surely, this set of rights is far from the set of property entitlements that we now enjoy, under which we may claim rights to luxury goods that cannot be consumed within our lifetimes. For Locke, this later, more extensive, set of property rights may be justified if systems of non-spoilage exist so that the spoilage condition can be met regardless of the amount of goods that we amass. Locke attributes the fulfilment of this condition to the invention of money, which doesn't spoil (Locke (2003[1690]), Chapter 5). We may amass money, more money than we may use within our lifetimes, because the value of money will not diminish; it is not perishable in the way that most natural products are perishable. So it is possible to understand the principle of acquisition in ways that justify weighty and extensive entitlements. The relevant question then is: can Locke's reasoning be adequately applied to territorial rights? May collectives indeed acquire territorial rights over as much area as they can 'amass', or are they more limited to an acquisition of an area that they can make 'use' of?

It seems that territorial rights are limited to the initial stage of rights, whereby collectives are only justified in acquiring what it is that they can 'use'. We should understand how we 'use' territorial rights. Territorial rights are primarily justified by virtue of the establishment of a rule of law over persons and goods within a region – that this rule of law is imperative for providing individual basic needs. The way that collectives 'use' territory is in their provision of the rule of law. In this sense, the use of territory is dissimilar to the use of property. The function of territorial rights consists first and foremost in the exercise of jurisdictional authority; the function of property rights, by contrast, consists in a much broader range of activities. On this understanding, a collective is misusing its territory (is letting it 'spoil') when it fails to provide legitimate institutions for those that are living within the territory, and also when the institutions fail to provide for members' basic needs.

Following on this distinction between territory and property, Locke's striking reference to the invention of money does not seem relevant for territorial rights. Obviously, the invention of money, while a significant consideration in the non-spoilage of property, is not relevant to the spoilage of territory. Simply amassing and holding money does not ensure that there is less injustice or the provision of just institutions. Money can be substituted for goods in a system of property, because one need not 'do' anything in order to hold property. Territorial rights are different – having a territorial right requires that the right holder act on something that is not fungible: a particular geographical space where persons reside and require governmental structure.

In fact, territorial rights cannot be passively held like property rights in money. No substitute (like money) can be made for territory because of the very specific way that collectives 'use' territory. Territorial jurisdictional authority is over persons and resources in order to secure access to the objects of individual basic needs. Thus, with no substitute for the institutions and actions that construct a just rule of law, the current state of affairs holds that collectives are 'wasting goods' when they hold goods that they are not using to establish a site of justice. A site of justice is a geographical domain within which the space and resources of the domain are organized to establish at least minimally just institutions which provide for the basic needs of members. The acquisition of territorial rights from the commons is justified if the acquiring collective is necessary to establish a site of justice within that territory. Overall, the acquisition of territory is limited to the initial stage of rights over goods, under which collectives are only justified in acquiring what it is that they can actually 'use'.

With this understanding of the principle of acquisition of territorial rights, we can draw a number of conclusions regarding the general scope of territorial rights. First, on this view, territorial rights can only be acquired from the commons where the establishment of a site of justice is feasible – where a collective has the definite capacity to provide secure access to the objects of members' basic needs through the establishment of a rule of law. Implied by this is that territory cannot be established in areas where there are no persons (as 'legitimate rule of law' implicates persons to rule over), or where the establishment of territorial rights is excessively tangential to the establishment of jurisdictional authority. Claims to territorial authority in Antarctica are a clear example of claims that would violate this general principle.[7] No persons have a permanent residence in Antarctica, and the number of persons who reside there temporarily is so small and transnational that the provision of jurisdictional authority can be provided at a distance. In essence, rule over temporary residents of Antarctica would be similar to the jurisdictional authority over cruise ships in international waters. In practise, a collective need not claim territorial authority over the international water itself in order to provide sufficient jurisdictional assistance to the persons on board.

In principle, this reasoning can be applied to claims made over resources located deep underground and undersea. The territorial powers of Russia, Denmark and others currently assert *territorial claims* over the sea and to the seabed and what lies underneath. Technological advances and the increasing retreat of polar ice make possible the exploration and exploitation of the resources within that region. Exclusive claims to the area by state powers are becoming increasingly relevant as the benefits for these agents seem more likely. However, under the foregoing principle of territorial acquisition, these claims are unjustified. Because the provision of legitimate institutions for members is at most only tangentially related to control over these areas (especially to the area of the sea floor), the *collective cannot 'use' the area as a territory* and hence cannot establish these institutions of jurisdictional authority.

In a similar vein, resources located deep underground should not fall within the territorial domain of a collective. At best, claims to resources located in the ocean and deep underground are property claims. Yet these property claims should be regulated by an international body capable of regulating *commonly* held goods, as will be explained below into more detail.

Certain indigenous groups, such as the Inuit and the Sami, maintain trade, are hunting and protect herding routes over the Arctic region that are integrally vital to their traditional way of life. Evidently, these cases may count as exceptions to the arguments I have been making here. Where a people resides within a region and the resources of the region form an integral part of meeting their basic needs, then that territory may definitely be claimed as a region of jurisdictional authority by that people. In these cases, the people meets a necessary condition of claiming territorial rights, since that jurisdictional control over the region is absolutely necessary to provide for the basic needs of its members. However, this claim is by itself not sufficient. The people also has to constitute a group with a legitimate claim to self-determination. In several cases, such as the case of the residents of the Falklands, the group may be just too small and possess too few economic resources to support a viable, independent state. In these cases, the group may claim embedded territorial rights under the auspices of a larger state authority.

Of course, an objection could be made that the development and use of resources within these remote regions could be used to meet the needs of the collective's members. Russia may use the proceeds from the sale of resources under the seabed of the North Pole to establish basic medical facilities for its members, for example. This use of resources is commendable, but it is not sufficient to meet the conditions of acquiring territorial rights. In this case Russia would not be *using* this region as a site of justice. Rather, Russia may possibly acquire a *property* right to those resources and use them for this end. Collectives may hold property rights in goods that are not within the territorial domain. This situation is similar to any foreign person or agent holding a property right within another state's territorial domain. The property right can be held by the foreign agent as long as the foreigner observes the law of the domestic state (regarding that specific property right).

In the case of property held in areas where there exists no domestic territorial authority, the property should be considered to be under the domain of the international community. UNCLOS (the United Nations Convention on the Law of the Sea) and the UN Division of Ocean Affairs serve as a legislative and weak enforcement agency for the management of rights over the high seas. While not meeting the conditions of a legitimate self-determining collective, the international community has developed a minimal system of adjudicating and articulating ownership rights that allows for effective administration of property rights in these relatively unpopulated areas. Principles regarding the legitimate management of property rights by an international (or supra-national) authority such as the just mentioned UNCLOS and the UN Division of Ocean Affairs could be derived from universal principles, such as an equal concern for meeting the basic needs of each individual, and from legitimate international treaties. As a matter of fact, the arguments here could be used to justify the extension of the authority of UNCLOS to Antarctica and to deep underground resources.

A Possible Objection

In the context of resource scarcity and improved access to Arctic and underground resources, one may also take up the argument that we have a duty to acquire resources from the commons in order to prevent a tragedy of the commons. David Schmidtz argues that privately appropriating goods from the commons is justified and sometimes required if the appropriation prevents the destruction of the commons. The tragedy of the commons is the result of a collective action that occurs when a group of individuals are each individually motivated to take more from the commons than the commons can sustain. Schmidtz illustrates this point with apples:

> A group of us find ourselves instantiated in a valley full of various kinds of apple trees, some already bearing ripe fruit. Finding ourselves hungry, we commence picking and eating apples. But some apples are better than others, so as other trees begin to ripen, a few of us jump the gun and pick the best apples before they are ripe, and the rest of us rush in, fearing that we will otherwise lose out entirely. Things get worse and eventually we find ourselves picking fruit before it is really edible. ... As latecomers arrive and natural resources become scarce, anything left in the commons will be levelled. ... leaving goods in the commons practically ensures their ruin. (Schmidtz, 1990)

For Schmidtz, we are able to provide ongoing access to the resource that we need (apples) if individuals appropriate land. Because the privatization of land restricts access to apples, a tragedy of the commons regarding apples is avoided when land is appropriated. The persons who privately own land are motivated to take care of their apple trees in order to exchange apples for other 'goods', such as money or labour. The persons who do not own land still have access to apples through the exchange of their money or labour for apples. Thus the privatization of land is crucial for all persons to have access to the objects of their basic needs. Even if, as Elinor Ostrom suggests, collective management of resources best avoids the tragedy of the commons in certain circumstances, collective management still demands a defined collective with the capacity to communicate and regulate each other within defined boundaries (Ostrom, 1990). Under this Ostrom view, the resources are 'taken from the commons' in that they are under the control of a particular group and not left to random taking by others.

Schmidtz's analysis can be adapted to resources which are currently left in the de facto commons. By de facto commons, I mean those resources which meet the conditions of being in the commons, regardless of their supposed legal status. On most accounts of 'the commons', a resource is in the commons if (1) access to the resources is difficult to regulate and if (2) the negative effects of its use are dispersed across a broad range of agents. In Schmidtz's case, the apples are taken out of the commons when the land containing the apples is appropriated. As such, land privatization allows the owner to restrict access to the apples and constrains the negative effects of misusing the apples to the land owner herself. The Arctic and Antarctic regions are in the commons because they are vast, largely unoccupied regions. Any agent with aquatic and cold-weather technology can access them. When more agents possess this technology and when these resources (fresh water, oil and gas, minerals and so on) become more precious, the incentive to exploit them will be greater. Soon these regions will look much like Schmidtz's common apple grove; the resources may be destroyed by the unconstrained pursuit of individual ends. On this reasoning, these regions ought to be appropriated by agents, such as states or international organizations, with the capacity to secure these large regions, before the conditions motivating the destruction of the commons take hold.

In general terms, however, the appropriation of the commons in the case of the Arctic and Antarctic regions is not relevantly analogous to Schmidtz's apple grove, and the appropriation of these regions may do more to bring on the destruction of the commons than to prevent it. To see why this is the case, it is useful to see why the appropriation of deep underground resources is no guarantee to stop the tragedy of the commons. First, deep underground oil wells are not renewable. Apples, however, are renewable. The appropriation of land with apple groves incentivizes the owner to care for the apple trees so that the value of the orchard and apples will be sustainable in the future. The value of underground wells, however, is not renewable. This resource only allows for one-time use and will then be destroyed. The owner of land containing underground oil wells has an incentive to extract the oil as cheaply as possible for a one-run big payoff. In the case of non-renewable resources, the owner is not motivated to care for the land, because its primary value is not renewable.

Second, in the case of the apple grove, the land is utilizable for many purposes. If the owner decides to herd sheep or operate a bed and breakfast instead of growing apples, the owner may cultivate the land for these purposes instead of for an orchard. For instance in the case of deep underground resources, the owner may not own any surface land, or only the amount of land that is necessary for the extraction of the resource. The underground area is not utilizable for other purposes – at least not in any accessible sense. The underground area is used for only one thing: the extraction of the resource. Once the resource is gone, the area underground is no longer of use.

Again, there is no incentive for the individual to tend towards the preservation of the resource or the area underground, because the primary value of the privatized region is the non-renewable resource. Generally, private owners are motivated to extract the resource for a big one-time payoff, and then move on to another investment. In this sense, the privatization of deep underground resources does not solve the tragedy of the commons. In many cases, the individual is still rationally motivated to engage in activities that destroy the common resource. Alternatively, if the area is used as a site of justice, it has many purposes which would encourage sustainable maintenance. However in these cases where a site of justice is impossible, the destruction of the commons is likely because the key social features that encourage appropriate resource management and accountability are missing.

The same analysis holds for Antarctic and Arctic regions. The resources there are like non-renewable apples in the commons. Because the land is not accessibly utilizable for other reasons, the agents are motivated to take all that is available before somebody else gets it first.

Generally, resources of a certain kind (that are (a) not renewable and (b) their site has very limited uses) are better off remaining in the commons than in being fenced off by profit-seeking agents or by states. However, it is often overlooked that there exist several different ways that a resource can be left 'in common'. It seems clear that merely leaving it in a 'negative commons', the commons of the 'state of nature', where any agent has the liberty to (mis)use the resources, would invite destruction of the resources through the tragedy of the commons and would invite conflict over control of resources. Thus the best interpretation of the commons should be a version of the 'positive commons', where each individual has an ownership share of the natural resource. In this case, each individual is a shareholder in the resources that are left in the commons. This means that rights over common resources should be adjudicated through a global institution that approximates global democratic decisions. While these institutions are far from being institutionalized currently, the UNCLOS is an approximation that may serve as an intermediary institution for this purpose.

From the above we can establish a line of defense against another often mentioned objection. Certain theories of territory make it possible for groups to have claims to territory which they do not inhabit. For example, geographical regions may fall under territorial rights if a right to that region is necessary to secure cultural value (on nationalist views) or projected diverse value (on Avery Kolers' plentitude view) (Kolers, 2009). On both of these kinds of theories, a group may find significant value in a region that they do not inhabit, and this value can ground a claim to territorial rights. Perhaps Russia could claim the Arctic on symbolic cultural grounds. However, given the above analysis that acquisition of certain kinds of resources destroys the commons, both of these kinds of theories would fail to justify taking of goods from the commons because that acquisition (a) would not serve the basic needs of individuals (as they would not establish a site of justice) and (b) would contribute to the failure of individuals to meet their basic needs. Hence, the principle of basic needs both establishes a rational ground for the acquisition of territory from the commons and at the same time limits the scope of territorial claims, even if those territorial claims are made on alternative normative grounds.

CONCLUSION

I have argued that a collective is justified in acquiring territory from the commons if that acquisition is necessary to establish a site of justice for the collective's members. On this view, the scope of territorial rights is strictly limited to only those areas where a collective can establish a site of justice sustaining just institutions. This limitation in scope excludes territorial claims of, for instance, Russia, Denmark and Canada over Arctic regions and deep underground resources. Although these resources cannot be appropriated from the commons by territorial claims, they should not be left in the state of nature – where any agent has a liberty to exploit them. Preferably, they should be considered as an integral part of a commons. Because the good management of these resources is essential to the meeting of individual basic needs, we should consider these common resources to be held in a *positive* commons ... where each individual is a shareholder in the resource. While the articulation of global democratic institutions capable of administering these common resources is beyond the scope of this chapter, I have indicated that an intermediary institution and appropriate first model may be found in the UNCLOS and the UN Division of Ocean Affairs.

NOTES

1. See John Locke (2003[1690]), sections 116, 117. Elsewhere I argue that Locke's account of territorial rights has contradictory results. Moreover, attempts to explain territorial rights through pre-established property rights fail. See Nine (2012, Chapter 2) and Nine (2008a, 2008b).
2. See Raz (1986). Collective rights are rights that individuals hold collectively but that individuals cannot hold individually, and a collective right is justified on the basis of individual interests.
3. For a good discussion in the context of global justice, see Brock (2009).
4. In order for persons to have control over their environment, the political structures must be organized such that they have the opportunity to express opinions and influence political decisions. For our purposes, we can interpret 'political control' as, first, the ability to express and form one's own opinions, and second, that a significant set of political decisions affecting one's life reasonably reflect these opinions. Meeting these conditions does not require that a person is able to directly and equally participate in political decisions.
5. See John Locke (2003[1690]). Similar positions are taken by Immanuel Kant and others: state oversight is necessary to legitimate and stabilize property rights.
6. This is consistent with certain contemporary theories of state legitimacy (see Thomas Pogge, 2002; Allen Buchanan, 2004; Martha Nussbaum, 2000).
7. Currently there are eight different territorial claims made to areas within Antarctica.

REFERENCES

Brock, G. (2009). *Global justice: A cosmopolitan account* (p. 66). Oxford: Oxford University Press.

Buchanan, A. (2004). *Justice, Legitimacy and Self-Determination*. Oxford: Oxford University Press.

Buckle, S. (2002). *Natural law and the theory of property: Grotius to Hume* (pp. 95–96). New York, NY: Clarendon Oxford University Press.
de Soto, H. (2000). *The mystery of capital*. New York, NY: Basic Books.
Gans, C. (2001). Historical rights: The evaluation of nationalist claims to sovereignty. *Political Theory*, *29*, 58–79 (p. 61).
Grotius, H. (1625). *The law of war and peace* (F. W. Kelsey, Trans.). [n.p.]. Longang Institute. (pp. Book 2, Chapter 2).
Grotius, H. (1868). *De iure praedae commentarius* (*Commentary on the Law of Prize and Booty*). In G. Hamaker (Ed.), The Hague: Nijhoff.
Kolers. (2009). *Land conflict an d justice*. Cambridge: Cambridge University Press.
Locke, J. (2003[1690]). *Second treatise of civil government*. Cambridge: Cambridge University Press.
Meinzen-Dick, R. (2009). *Property rights for poverty reduction?* (p. 1). Report. Cambridge: Cambridge University Press.
Nine, C. (2008a). A Lockean theory of territory. *Political Studies*, *56*(1), 148–165.
Nine, C. (2008b). Territory is not derived from property: A response to Steiner. *Political Studies*, *56*(4), 957–963.
Nine, C. (2012). *Global justice and territory*. Oxford: Oxford University Press.
Nussbaum, M. (2000). *Women and human development* (p. 80). Cambridge: Cambridge University Press.
Ostrom, E. (1990). *Governing the commons: The evolution of institutions for collective action*. Cambridge: Cambridge University Press.
Pogge, T. (2002). *World poverty and human rights*. Cambridge, MA: Polity.
Raz, J. (1986). *The morality of freedom* (p. 180). Oxford: Clarendon Paperbacks.
Schmidtz, D. (1990). When is original appropriation required? *Monist*, *73*, 504–519.
Sen, A. (1980). *Equality of what?* Cambridge: Cambridge University Press.
Simmons, A. J. (2001). On the territorial rights of states. *Philosophical Issues: Social, Political and Legal Philosophy*, *11*, 300–326.

MANDATORY NON-ANTHROPOCENTRISM: THE POLITICAL UNREALISM OF MAKING METAETHICAL DEMANDS IN ENVIRONMENTAL ETHICS

Jason Dockstader

ABSTRACT

Purpose – *This chapter argues that environmental ethicists commit a serious error when they require that people hold a moral realist metaethical belief in the intrinsic value of non-human living things and non-living natural things in order to be able to behave in an ethically acceptable manner toward the environment.*

Methodology – *Environmental ethics regard this position as the mandatory non-anthropocentrism one must first hold in order to be in a proper moral relationship to the environment. The main reason for seeing this requirement as an error is that it is politically unrealistic insofar most people most of the time behave in political contexts on the basis of*

Environmental Philosophy: The Art of Life in a World of Limits
Advances in Sustainability and Environmental Justice, Volume 13, 157–173

ISSN: 2051-5030/doi:10.1108/S2051-5030(2013)0000013011

instrumental and not intrinsic reasons. To claim that people can behave in a morally acceptable manner toward the environment if and only if they first believe in its intrinsic value is not only politically unrealistic, but also actually false.

Findings – *The chapter looks at recent studies measuring the behavior of political and moral philosophers which shows that they do not behave in any markedly way better than non-moral philosophers. Ethicists, whom one can assume believe in some form or another of the mind-independent reality of moral properties, are not more morally well-behaved for holding such a belief.*

Implications – *Ethicists, especially environmental ethicists, are in no position to require of us to believe in the intrinsic value of the environment in order to behave in more beneficial ways toward it.*

Keywords: Anthropocentrism; intrinsic; environment; realism; politics; metaethics

My arguments in this chapter begin as reflections upon the work of Andrew Brennan and Yeuk-Sze Lo. In particular, I am responding to some key points they make in their entry on environmental ethics for the *Stanford Encyclopedia of Philosophy* and their 2010 book, *Understanding Environmental Philosophy*. Echoing J. Baird Callicott's claim, that "how to discover intrinsic value in nature is the defining problem for environmental ethics" (Callicot, 1999), Brennan and Lo approach environmental ethics first and foremost from the perspective of the distinction between instrumental and intrinsic value. They begin by noting that it is often regarded as morally wrong for human beings to pollute, destroy, and over-consume the planet and its resources. They then ask if such human activities are morally wrong either because a sustainable environment is essential for the survival of human beings or because "the natural environment and/or its various contents have certain values in their own right so that these values ought to be respected and protected in any case?" (Brennan & Lo, 2008). These two options amount to the difference between instrumental and intrinsic value. Instrumental value is the value things have as useful means to certain ends (a sustainable environment being useful for our survival and thus having value), while intrinsic value is the value things have as ends in themselves, irrespective of their possible utility (the environment as something valuable in and of itself). So, for many people, the planet, and its resources, has mostly only instrumental value insofar as it is useful for the preservation and enhanced prosperity of its species. The assumption made by the vast majority of environmental ethicists is, on the other hand, that the planet does not contain merely instrumental value, but rather possesses intrinsic value as well. The question is then, why is it important for environment ethicists that the environment be regarded and recognized as containing intrinsic value? The answer appears to be that if something is discovered to possesses intrinsic value – that is, discovered to be good or morally real or valuable in and of itself regardless of its possible utility – then, so the thinking goes, that "generates a prima facie direct moral duty on the part of moral agents to protect it or at least refrain from damaging it."[1]

We already do something similar with human beings. Most of us would have a tough time admitting that our fellow humans are of *only* instrumental value to us. Most of us most of the time do indeed use other people as means for a variety of further ends, but we also like to believe that we do not view even the people to whom our relationship with them is based practically entirely upon utility that they do not also contain some other, more essential, degree of intrinsic value. In other words, we like to believe, especially today here in the West, that people contain more than just instrumental value and that they are intrinsically valuable. Why are humans intrinsically valuable, we may wonder? It seems to be simply because they are human beings. Human beings are intrinsically valuable for human beings. We believed we have discovered and recognized their, our own, inherent worth. Today we call the intrinsic value of humans their "human rights." These rights warrant, demand, and ultimately make mandatory a moral duty to protect or at least not harm human beings. How this duty is enacted and performed is through, as far as I can tell, either the law, through the legal codification of human rights, or through morally intensified interpersonal relationships that result from moral reflection and moral education. So we aim to protect the intrinsic value of human beings either by punishing the violators of human rights or by becoming more morally sound or acute through the process, sometimes practical and other times abstract, of training ourselves to think, act, behave, and live more like ideal moral agents. The moral reality, the intrinsic value, of human being has become for us if not axiomatic then at least politically and legally necessary.

Now, environmental ethicists don't have a problem at all with this, with recognizing and protecting the intrinsic value of human beings. They are all for it. What they have a problem with is not extending the recognition and protection of intrinsic value to other entities in the world, starting with, obviously, non-human animals, but extending the recognition and protection all the way past non-living natural things like rivers and mountains all the way up to the entire planet as such, including all the life, matter, and energy within it.[2] The way that many humans – especially many moral philosophers throughout history, in both the East and the West – have ascribed intrinsic value to human beings alone is what environment ethicists call anthropocentrism, or human-centeredness. According to Katie McShane, anthropocentrism is the "assumption that human beings and/or their interests matter morally in their own right while everything else matters morally as it affects human beings and/or their interests" (McShane, 2009). Brennan and Lo offer Aristotle's line from Book I of the *Politics*, that "nature has made all things specifically for the sake of man," as a prime example of anthropocentrism. Kant, as well, took an anthropocentric view when, in his lecture "Duties to Animals and Spirits," he argued that beating a dog would be bad because it might lead a person to become cruel toward humans, not because a dog has intrinsic value that demands recognition and protection and so beating it would just be inherently wrong.

Environmental ethicists often view Richard Routley's thought experiment of the last person as proving the wrongness and badness of anthropocentrism. Routley's experiment goes like this: imagine some catastrophe strikes and every human on the planet dies except you. You are the last human alive, but you are dying as well. You are about to die. Now imagine that right before you die you can push a button that will kill the rest of life on the planet with you. As you die, you can make it such that all of life perishes along with you. As you go, everything else goes too. Would this act be morally wrong? Would it be wrong to end all life on the planet just because you were the last human and so humanity would no longer figure in the scheme of life? That is, would there be anything wrong with correlating your own demise with the obliteration of the entirety of life on the planet? One would think there really shouldn't be a problem if only humans contain intrinsic value. The environmental ethicist's response to these questions is to say "no way," most would not kill all the life on the planet even if the entire human species would go extinct with one's own death. Why is that so? Because, according to them, deep down, most of us believe non-human life contains some degree of intrinsic value, that is, most us are truly non-anthropocentric in some way. It is just not in us to murder all of life only because humans will no longer be around. If that is the case, then we should bring this revelation back to the real world and start behaving in accordance with this latent belief we supposedly have always already had. We now apparently have no reason whatsoever not to believe that other entities, in fact most if not all non-human living and natural entities, contain the intrinsic value we prior thought only belonged to ourselves. We should act in accordance with our own beliefs, on pain of being irrational and contradicting ourselves. We all know innately somehow that anthropocentrism leads to environmental damaging behavior. The inculcation of this conclusion into as many people as possible has been the goal of environmental ethics since its inception. As Brennan and Lo note, environmental ethics can almost be defined as the war against anthropocentrism:

> When environmental ethics emerged as a new sub-discipline of philosophy in the early 1970s, it did so by posing a challenge to traditional anthropocentrism. In the first place, it questioned the assumed moral superiority of human beings to members of other species on earth. In the second place, it investigated the possibility of rational arguments for assigning intrinsic value to the natural environment and its nonhuman contents. (Brennan & Lo, 2008)

But, how have environmental ethicists gone about doing this? How have they argued against anthropocentrism? Even more importantly, what did they think would be the effect of dropping an anthropocentric view on things and picking up the belief that non-human living things and non-living natural things possess intrinsic value? What was their strategy for inducing a change in people's behavior? To answer these questions, we have to look at some of the many environmentalisms that have emerged over the past 40 years. Before we do that though, we should look at the very convenient summary offered by Brennan and Lo of the logical structure of just about every environmental ethicist's argument against anthropocentrism. The argument always first establishes something as the cause of anthropocentrism, usually in the form of "X leads to or causes anthropocentrism." Next, it is asserted that anthropocentrism leads to environmentally damaging behaviors. And, finally, it is concluded that X, which leads to anthropocentrism, leads to or causes environmentally damaging behavior. X is thus the cause and origin of the environmental crisis. This argument is both philosophical and cultural-historical in that it combines the belief in the environment's intrinsic value that anthropocentrism ignores or denies with actual cultural-historical instantiations of anthropocentrism in practice.

To look at a few examples, let's start with Lynn White Jr's foundational 1967 article, "The Historical Roots of our Ecological Crisis." White argues that Judeo-Christian monotheism is the historical and ideological source of the modern environmental crisis. As Brennan and Lo summarize White, he maintained that "the main strands of Judeo-Christian thinking had encouraged the overexploitation of nature by maintaining the superiority of humans over all other forms of life on earth, and by depicting all of nature as created for the use of humans" (Brennan & Lo, 2008). It was within this Judeo-Christian context that modern Western science and technology could comfortably develop and expand its unlimited drive to exploit nature. Another example comes from ecofeminism. For many ecofeminists, like Val Plumwood, the historical and ideological source of the environmental crisis is "patriarchy and its radical separation of male and female into two evaluatively opposite spheres" (Brennan & Lo, 2010), with the male sphere usually being regarded as superior. The problem with patriarchy is that it prejudicially associates with masculinity the rational, active, creative, civilized, orderly, and transcendent aspect of human existence, thus associating with femininity all that is emotional, passive, determined, animalistic, primitive, disorderly, natural, physical, and immanent. The androcentrism (male-centeredness) of patriarchy clearly leads to anthropocentrism for ecofeminists. Therefore, patriarchy is ultimately the cause of the environmental crisis. Another example comes from the deep ecological movement. According to deep ecologists, like Arne Naess, what really leads to the anthropocentrism that causes the environmental crisis is the belief in atomistic individualism, the belief that human beings are individuals in the first place. By taking a relational and holistic approach to the natural world, deep ecologists believe that environmental degradation stems from the unjustified belief in the substantially individuated distinctness of human beings over and against the natural world. Humans are really just more knots in the whole immense web of the interconnected interrelatedness of nature. For deep ecology, we need to extend out into our true ecological Self and so overcome our all-too-human (or perhaps just Western) view that we are actually distinct individuals. A final example comes from a group of thinkers sometimes called new animists, who often also self-identify as panpsychists, but which can also include critical theorists like Adorno and Horkheimer and even aspects of Max Weber's social theory. For this group, what leads to environmentally damaging anthropocentrism is the general disenchantment of nature. Through the positivistic rationalizing and bureaucratizing techniques of capitalist exploitation of all forms of life and nature, the universe becomes a meaningless and soulless place, a truly disenchanted world. From this perspective, nature is no longer mysterious, no longer sacred. Devoid of any threatening or even interesting opacity, nature not only becomes a place wherein it is impossible to experience the dread and awe, the wonder and sublimity, that once made life worth living, but also becomes nothing more than a giant machine whose inner secrets have been laid bare for the sake of a maximally mechanistic manipulation.

What each of these environmental ethics share is the belief that something necessarily leads to anthropocentrism, which necessarily leads to environmental degradation. That something or X is Judeo-Christianity, patriarchy, individualism, or disenchantment for Lynn White Jr, ecofeminism, deep ecology, and new animism, respectively. Implicit to each of their arguments is the fundamental belief in the intrinsic value, the moral reality, of the natural world and the life in it. Brennan and Lo call this implicit belief the evaluative thesis, which is "the claim that natural nonhuman things have intrinsic value, i.e., value in their own right independent of any use they have for others" (Brennan & Lo, 2008). The evaluative thesis, they claim, "has now achieved a nearly canonical status within the discipline" (Brennan & Lo, 2008). Along with the evaluative thesis, Brennan and Lo see another thesis as being implicit to most environmental ethics, a thesis based on the assumption that non-anthropocentrism is the solution to anthropocentrism and that environmental problems can be solved if and only if people start to believe in, abide by, and behave in accordance with the evaluative thesis. This is the psycho-behavioral thesis, which is "the claim that people who believe in the evaluative thesis of non-anthropocentrism are more likely to behave environmentally (i.e., behave in beneficial ways, or at least not in harmful ways, toward the environment) than those who do not" (Brennan & Lo, 2008). For Brennan and Lo, the psycho-behavioral thesis is important because it "provides a rationale for both the diagnosis and solution of environmental problems, and it gives practical justification to the discipline of environmental ethics itself (conceived as the mission to secure converts to the evaluative thesis of non-anthropocentrism)" (Brennan & Lo, 2008). Now, I think there are many problems with the psycho-behavioral thesis and also, by extension, the evaluative thesis, but I will focus more on the former. As Brennan and Lo note, if the psycho-behavioral thesis turns out to be false, then the diagnoses and solutions offered by environmental ethics, and environmental ethics itself as a sub-discipline of applied ethics, may be seriously undermined and rendered futile, and I am seriously afraid that that is the conclusion I reach in this chapter.

First of all, let's take a step back and look at how the discovery, recognition, and protection of the intrinsic value of human beings is supposed to allow for the same discovery, recognition, and protection to be extended out to other living things and eventually the environment itself. Most ethics, normative and applied, begin with the assumption that humans, just by being alive, are of some value in and of themselves. It has been notoriously difficult to justify this assumption on secular grounds, however. If human beings are no longer inherently worthy because some transcendent deity loves them all equally and so endows them certain inalienable rights, then there will have to be a different source for their intrinsic value. Usually, that source has something to do with the fact that we feel we are conscious beings with some degree of freedom of choice or, looking at Article I of the UN Universal Declaration of Human Rights, that we are all "born free and equal in dignity and rights" because we are "endowed with reason and conscience." The special value of humans consists in their conscious and conscientious use of their freedom and reason. It is exactly at this point that many environmentalists and especially animal rights activists point out that many other, if not all, animals exhibit extremely similar if not identical features. The expanding circles of recognized intrinsic value can now begin to ripple, unless we admit to being "speciesist," to use Peter Singer's term, arguing that only humans are special. But, on the face of it, who wants to be a confessed speciest?

Interestingly, at this point, the wider the net is cast and the more it looks like a greater variety of other species exhibit the same properties that made humans uniquely special, *the less humans look uniquely special.* Humans do not specifically contain intrinsic value, we are not special; rather, intrinsic value based on some sort of conscious and conscientious reasoning capacity is a fairly common property found in many other living things. But this is a problem because the starting point of the entire process of expanding the recognition of intrinsic value now appears quite dubious as a starting point at all precisely because humans are not special in their possession of intrinsic value. If humans are not especially or specifically intrinsically valuable, how can we start with the specific value of humans as that point from which we can expand the circle of the recognized and protected intrinsic value of other living things? I can think of only a couple of solutions to this problem, but environmental ethicists will not like them. Maybe we can rein in our expansionist desires and just live with the fact that we are in some basic sense anthropocentric or speciest, believing as we do in only our own intrinsic value; or perhaps at least we can admit that the entire global economy and thus the security and comfort of living in the rich, developed world depends upon the large-scale destruction of many animals, plants, resources, and ecosystems and that we are not so willing to be non-anthropocentric to the point that it would disrupt our own lives. Bryan Norton has made a roughly similar argument in certain works. McShane summarizes his position by writing that "environmental ethicists should accept anthropocentrism and give up the search for nonanthropocentric theories of value" because "many policy-makers as well as social scientists whose views affect environmental policy (e.g. economists) assume the truth of anthropocentrism" (McShane, 2009). Bernard Williams has also made similar points in his essay, "Must a Concern for the Environment be Centered on Human Beings?" His answer is yes because he finds it impossible that humans could ever somehow escape their own very limited, subjective, finite, and fallible perspectives on the world and so discover what is objectively or intrinsically valuable within it. He also notes, I think importantly, that there are some living things that it would be downright suicidal for us to try to recognize and protect, like cancer and AIDS and other horrible diseases with which we are in an eternal fight to the death. I still find astounding some of those extraordinarily absurd moments in environmental ethics when someone, like Paul Taylor with his idea of "biocentric egalitarianism," asserts that we have a duty to protect or at least not harm every single living thing, each of which Taylor calls a "teleological center of life," because they are imbued with "inherent worth" and "well-being." The employment of such a deontological approach to environmental ethics amounts to little more than a death wish. It is hard to deny that there are some living things out there that do not quite care for our recognition of their intrinsic value and will most definitely kill us if we do not kill or at least contain them first.

Now, this option of admitted speciesism may be acceptable to a lot of people (it is to my father, who is the most average person I know), but I do not think most environmental or any other kind of ethicist would accept it. Maybe, on the other hand, we can just admit that intrinsic value or inherent worth, the perceived essential moral reality of things in the world independent of our minds, is ultimately an inexplicable and brute fact, a necessary yet convenient mystery, or a post-hoc rationalization of our experiences of pleasure and pain, or an unreflective intuitive projection of our affects onto the world, or an instinctual expression that takes the form of certain assertions and propositions that are fundamentally yet usefully false or that are least subjectively dependent upon our own particularly conative existence? Maybe we could just admit that the belief in the intrinsic value of anything, especially humans, is just a beneficial fiction we could no longer live without? Again, I do not think any environmental or any other kind of applied or normative ethicist would accept any of these options because they are all metaethical moral realists, that is, they believe there are moral facts about the world, like intrinsic value, that can be known and which can and must guide our behavior. Another problem they would have with the above options, all of which entertain taking a metaethical moral anti-realist view, is that they tend to provide explanations for the belief in intrinsic value in non-intrinsic, that is, instrumental, terms. Even if there may be evolutionary explanations as to how becoming increasingly morally realist in our beliefs is particularly beneficial for the survival and thriving of human beings and even other species, they want to hear nothing of it because it always smacks of instrumentalist and anthropocentric reasoning to them. Discovering, recognizing, and protecting intrinsic value – or to speak in more anti-realist terms: imagining, projecting, and prejudicially believing in intrinsic value – must itself be intrinsically valuable for them. It seems then that blatantly violating the Principle of Sufficient Reason is of not much concern to environmental ethicists. I think this is because they believe they have bigger fish to fry. They have behavior to change!

As we saw, the psycho-behavioral thesis is the key implicit assumption of all environmental ethics. The assumption is that by viewing the world non-anthropocentrically one tends to behave in more environmentally helpful ways. By losing a Judeo-Christian worldview or becoming an ecofeminist, deep ecologist, or new animist, one can thereby not only become a better environmental activist, but also can actually come to practice him or her everyday the recognition and protection, the intrinsic value of non-human living, and natural things. How about a simple question: are they right? Is it true that by believing in the intrinsic value or moral reality of things one behaves more morally? This is an empirical question that can only be answered empirically. In fact, the initial points of each of our representative environmental ethicists were also empirical. Are they right to say, historically or culturally speaking, that it is a fact that Judeo-Christianity, patriarchy, individualism, or positivism are the causes of environmentally damaging behavior? Only an empirical, historical analysis could answer such a question. And only could an empirical study discover whether or not the psycho-behavioral thesis holds. No a priori philosophical reasoning could help here. The problem is that so far there have only been a few studies that have attempted to answer these questions. Brennan and Lo write that "of the major philosophical theories on the origin of environmental crisis, Lynn White's is the only one to has been empirically tested by social scientists" and that "the net result of these studies so far has been "inconclusive," especially when education, sex, age and social class are also factored in" (Brennan & Lo, 2008). Another study mentioned by Brennan and Lo is on agricultural practices in Latin America by William Durham, which "showed that families who held animist views about nature, and who were careful to ask forgiveness for land clearing that disrupted indigenous vegetation and animals, had a *worse effect* on the environment than families who aimed to maximize their income from farming and had no particular conception of the land as intrinsically valuable or sacred. It turned out that the economically oriented families had ecologically superior land rotation practices" (Brennan & Lo, 2008). These results lead one to observe that the presuppositions and expectations of environmental ethicists have yet to be proved, that the results in fact tend toward disproving them, and that, even more, it is intellectually and concretely hazardous to assume certain unsubstantiated historical facts hold and certain people's beliefs and values about these facts necessarily translate into actions and behaviors which necessarily follow from them. The transition from belief to action is highly precarious and rarely works out in such a way that there is clear correspondence or identity between the two. This is what social and moral psychologists have been trying to tell us for at least the past 30 years.

Keep in mind that the environmental ethicist's assumption is that people who believe in intrinsic value will tend to be the kind of people who behave, one would think consistently, in accordance with their belief. These people will possess certain robust psychological and character traits that will enable them to actualize their beliefs in practice. Believers in intrinsic value *will* be better environmentalists, the thesis presumes. Social and moral psychologists, however, like John Doris, Stephen Stich, and Gilbert Harman have shown that this is rarely the case. They refer to the notion of situationism, which holds that people are more influenced in their actions by external, situational factors than internal, motivational, or cognitive factors, like beliefs or supposed character traits. Doris and Stich write that

> social psychologists have repeatedly found that the difference between good conduct and bad appears to reside in the situation more than the person; both disappointing omissions and appalling actions are readily induced through seemingly minor situational features. What makes these findings so striking is just how insubstantial the situational influences effecting troubling moral failures seem to be; it is not that people fail to adhere to standards for good conduct [I think we can read here 'belief in intrinsic value' as a standard, if not *the* standard, for good conduct], but that they can be induced to do so with such ease. (Doris & Stich, 2006)

Doris and Stich provide a summary of some classic studies in the field from the 1970s as examples:

> Isen and Levin (1972: 387) discovered that subjects who had just found a dime were 22 times more likely to help a woman who had dropped some papers than subjects who did not find a dime (88% v. 4%). Darley and Batson (1973: 105) report that passersby

> not in a hurry were 6 times more likely to help an unfortunate who appeared to be in significant distress than were passersby in a hurry (63% v. 10%). Mathews and Canon (1975: 574–5) found subjects were 5 times more likely to help an apparently injured man who had dropped some books when ambient noise was at normal levels than when a power lawnmower was running nearby (80% v. 15%). Haney *et al.* (1973) describe how college students role-playing as "guards" in a simulated prison subjected student "prisoners" to intense verbal and emotional abuse. Milgram (1974) found that subjects would repeatedly "punish" a screaming "victim" with realistic (but simulated) electric shocks at the polite request of an experimenter. (Brennan & Lo, 2008)

Even though I have not found many studies that similarly try to test people's environmental behavior in different circumstances which involve different situational factors, I find it hard to believe that we wouldn't also find that in the majority of the cases people's supposed environmentally sound behavior could be very easily curtailed and checked by the most minute and irrelevant of influences. It appears then that what the social science tells us is that believing in the intrinsic value of the environment would in no way guarantee or even render probable that people would behave more environmentally sound. Belief in intrinsic value may only barely translate into correspondent action. I think this is probably because people are often wildly inconsistent with respect to practicing their beliefs, which is to say, in other words, maybe more positively, that people are only barely locally consistent in actualizing and practicing their beliefs or remaining characteristically regular. This is not to blame people for failing or to assume that one could even get away with perfectly actualizing a belief in practice. It may be rather that since beliefs are for the most part incredibly vague dispositional tendencies pushing us in a variety of different directions at once – I mean, let's admit, most of our beliefs are half-baked in every sense of that term – that neatly actualizing beliefs, any beliefs, in practice may be relatively impossible.

Let's look at this from one more angle: if believing in the intrinsic value or moral reality of things is supposed to lead one to behave more morally, then wouldn't the very people, the ethicists, who come up with these ideas behave more morally themselves? That is, wouldn't ethicists and moral and political philosophers who believe in intrinsic value behave better than the rest of us? Eric Schwitzgebel and Joshua Rust have pretty much asked this very question and done some studies to answer it. In the abstract to their study, "The Self-Reported Moral Behavior of Ethics Professors," Schwitzgebel and Rust summarize their approach in this way:

> We examine the self-reported moral attitudes and moral behavior of 198 ethics professors, 208 non-ethicist philosophers, and 167 professors in departments other than

> philosophy on eight moral issues: academic society membership, voting, staying in touch with one's mother, vegetarianism, organ and blood donation, responsiveness to student emails, charitable giving, and honesty in responding to survey questionnaires. On some issues we also had direct behavioral measures that we could compare with self-report. Ethicists expressed somewhat more stringent normative attitudes on some issues, such as vegetarianism and charitable donation. However, on no issue did ethicists show significantly better behavior than the two comparison groups. (Schwitzgebel & Rust, 2011)

That is to say, Schwitzgebel and Rust found that ethics professors behave no better than non-ethics philosophy professors or other academics. This is contrary to the view most ethicists have of themselves, which is what Schwitzgebel and Rust call the "booster view." The booster view "predicts that ethicists will show the best moral behavior, the truest moral attitudes, and the strongest attitude–behavior relationship on the assumption that they tend to engage in more philosophical moral reflection than do the comparison groups and are otherwise socially similar in dimensions relevant to the moral behavior under investigation" (Brennan & Lo, 2008, p. 64). Schwitzgebel and Rust concluded:

> This model, we think, is very difficult to sustain in light of our data on moral behavior. We now have a number of direct measures of moral behavior. In previous research, we have found ethicists, including political philosophers, to vote at about the same rate as other professors (Schwitzgebel & Rust, 2010), to reply to emails apparently written by students at about the same rate as other professors (Rust and Schwitzgebel in preparation), to behave about as courteously at philosophy conferences (though environmental ethicists did appear to litter less; Schwitzgebel, Rust, Huang, Moore, and Coates forthcoming), to be rated morally about the same overall by their peers (Schwitzgebel & Rust, 2009), and to misappropriate library books no less often (perhaps more often, Schwitzgebel, 2009). In the current research, we found ethicists about as likely to pay membership dues to the APA, to be about as accurate or honest in their survey responses, and to be no more responsive than the other groups to the charity incentive (perhaps less responsive). Turning to self-report measures, all groups reported similarly high rates of having an organ donor symbol on their driver's license and similarly low rates of blood donation. Non-philosophers reported keeping best in contact with their mothers; non-philosophers and ethicists reported similar overall rates of charitable donation, somewhat higher than non-philosophers'; and non-philosophers were marginally most likely to report having eating meat at the previous evening meal. We regard these results as an approximate tie among the groups – as is also suggested by our three summary measures. (Brennan & Lo, 2008)

Again, a tie means that ethicists do not behave markedly different than non-ethicists. Now, if environmental ethicists are telling us that the more people believing in and acting morally in accordance with the intrinsic value of non-human things is the solution to environmental problems, then how could they possibly be right if there is no proof that ethicists, environmental ethicists included, themselves behave any better or differently than the rest of us? Moral and political philosophers don't even vote more than non-ethicists or non-philosophers (political science professors vote more), and voting is supposed to be the sacred moral duty of every citizen who believes in the normative legitimacy of democratic governance. This takes us to the final problem involved in arguing for both the evaluative and psycho-behavioral thesis: it is politically unrealistic.

Implicit to the evaluative and psycho-behavioral theses is the idea that the recognition of an ethical belief can and must be done *prior* to its enactment, that the right kind of politics can occur only *after* an ideal ethical theory has been established. What I mean is that for environmental ethicists the evaluative thesis must come first, one must recognize and appreciate intrinsic value first, and only then can one alter one's psychological view on the world and so change their behavior. The Cambridge political philosopher and self-described political realist, Raymond Geuss, has argued that this is exactly the kind of approach to political action that is incredibly unrealistic. Geuss begins his work *Philosophy and Real Politics* by addressing two distinct ways of understanding the phrase, "politics is applied ethics." The understanding of the phrase that Geuss finds problematic and unrealistic is the one he labels the "ethics-first view." The problem with the "ethics-first view" is that it approaches political action by first abstracting attention out and away from the spatio-temporal and embodied everyday lives of actual political actors so as to allow one to start thinking about the human social world through the process of trying to discover an ideal theory of ethical action. This search for an ideal theory rests on locating a general principle to guide the further abstraction that will eventually bring us back to reality, but only after we are morally ready and ethically prepared to act accordingly. Such a general principle usually sounds something like the phrase "humans are rational," or "people seek pleasure and aim to avoid pain," or even I think we could now add "non-human living and natural things contain intrinsic value." Next, more particular principles follow from whichever general principle has been chosen. For example, following the general principle of humanity's rationality the particular principle of striving to construct the ideal speech community or democracy or socialist paradise or whatever can emerge as a sort of guiding ideal or goal that can work as a constraint on action. In environmental ethics, following the general principle of intrinsic value, as we have seen, we find the particular injunction to recognize and protect it. So, finally, the thinking goes, after the work of abstracting out to and constructing an ideal ethical theory of political action based on general and particular principles is completed one can come back down to earth and start acting in accordance with these principles. Only now can one start behaving like a good environmentalist in one's political interactions. Geuss' problem with this entire approach is that it rests on the fundamental illusion that one can complete the work of ethics first and then go ahead and be a morally good political actor thereafter. As Geuss writes, "the view I am rejecting assumes that one can complete the work of ethics first, attaining an ideal theory of how we should act, and then in a second step, one can apply that ideal theory to the action of political agents" (Geuss, 2008). Geuss' point is that things simply do not work that way.

Now, there is another understanding of the phrase, "politics is applied ethics," that Geuss finds totally feasible. The anodyne or harmless understanding of the phrase Geuss finds acceptable I think actually deftly incorporates the findings of moral psychology mentioned above. Geuss writes:

> The anodyne reading asserts that 'politics' – meaning both forms of political action and ways of studying forms of political action – is not and cannot be a strictly value-free enterprise, and so is in the very general sense an 'ethical' activity. Politics is a matter of human, and not merely mechanical, interaction between individuals, institutions, or groups. ... Political actors are generally pursuing certain conceptions of the 'good,' and acting in the light of what they take to be permissible. This is true despite the undeniable fact that most human agents most of the time are weak, easily distracted, deeply conflicted, and confused, and that they therefore do not always do only things they take to be permissible. (Brennan & Lo, 2008, pp. 1, 2)

Geuss is not saying we cannot or should not strive to realize in political practice whatever conception of the good we may have – that is, in fact, just about all we can do in politics – but that we must take notice of the fact that "any attempt to think seriously about the relation between politics and ethics must remain cognitively sensitive to the fact that people's beliefs, values, desires, moral conceptions, etc., are usually ... both indeterminate and, to the extent to which they are determinate, grossly inconsistent in any but the most local, highly formalized contexts, and are constantly changing. None of this implies that it might not be of the utmost importance to aspire to ensure relative stability and consistency in certain limited domains" (Brennan & Lo, 2008, pp. 3, 4). I would say, and I think Geuss would agree, that environmental matters are certainly one of these domains. Real political action in the context of real political events and real political actors is much like trying to survive in a free-for-all. For the basic reason of human frailty, it is mostly impossible that the transition from the evaluative to the psycho-behavioral thesis could ever actually work in any context of real politics. It is politically unrealistic to expect a majority of people to both believe in the moral reality of non-human living and natural things and then to act in accordance with that belief.

To conclude, then, let me quickly summarize. I have made three arguments against what Brennan and Lo regard as the latent theses of all environmental ethics. First, I argued that there is no non-self-defeating way of defending the idea of either the intrinsic value of human beings or the intrinsic value of any other thing. This is more a metaethical point in favor of moral anti-realism, which, admittedly, no normative and applied ethicist would ever be willing to accept. Second, I argued that there is no conclusive proof either that the historical judgments in the evaluative thesis are correct or that the psycho-behavioral thesis can actually neatly follow from the evaluate thesis. Rather, there is evidence that moral beliefs or supposed robust psychological traits rarely if ever translate into consistent moral actions. Also, ethicists have been found to behave no differently than non-ethicists. Third, I argued that the idea of going first from the ideality of the evaluative thesis and then to the putatively less abstract practice of the psycho-behavioral thesis is politically unrealistic, that real politics does not allow for any concise application of ideal ethical principles.

If these arguments undermine the entire project environmental ethics, then it would seem environmental ethics has some work to do with respect to re-conceptualizing its foundational principles, even if its goals remain laudable and ultimately necessary.

NOTES

1. Brennan and Lo (2008) also refer here to the important work of John O'Neil and Dale Jameson on the issue of intrinsic value in environmental ethics.

2. To be honest, I can see no reason why a good environmental ethicist would have to stop there, at the planetary level. I mean, why not recognize and protect the intrinsic value of other planets, the Sun, our entire solar system, our galaxy, other galaxies (doesn't the Andromeda Galaxy contain intrinsic value?), and maybe even the entire universe from the Big Bang to the Big Crunch. Of course, I am being facetious, but the issue of the indefinite extrapolation of inherent worth is going to prove highly relevant in a moment.

REFERENCES

Brennan, A., & Lo, Y.-S. (2008). Environmental ethics. *The Standford Encyclopedia of Philosophy*. Retrieved from http://plato.stanford.edu/entries/ethics-environmental/

Brennan, A., & Lo, Y.-S. (2010). *Understanding environmental philosophy* (p. 166). Durham: Acumen.

Callicot, J. B. (1999). Intrinsic value in nature: A metaethical analysis, In *Beyond the land ethic: More essays in environmental philosophy* (p. 240). Albany, NY: SUNY Press.

Doris, J., & Stich, S. (2006). Moral psychology: Empirical approaches. *The Stanford Encyclopedia of Philosophy*. Retrieved from http://plato.stanford.edu/entries/moral-psych-emp/

Geuss, R. (2008). *Philosophy and real politics* (p. 8). Princeton, NJ: Princeton University Press.

McShane, K. (2009). Environmental ethics: An overview. *Philosophy Compass*, *4*(3), 407.

Schwitzgebel, E., & Rust, J. (2011). *The self-reported moral behavior of ethics professors*, p. 1. Retrieved from http://www.faculty.ucr.edu/~eschwitz/SchwitzPapers/EthSelfRep-110316.pdf

IRISH RUINS ANCIENT AND NEW: GHOST ESTATES, MEGALITHS AND HUMAN RELATIONS WITH THE REST OF NATURE

Geoff Berry

ABSTRACT

Purpose – *This chapter considers the environmental damage related to Ireland's recent 'ghost estates', placing this disastrous waste of resources in the long historical context of ancient ruins that also dot the land.*

Methodology/approach – *It considers ruins from an ecocritical perspective, as material artefacts attesting directly to people's relationship with their environment.*

Findings – *From ancient megaliths and sacred sites to imposing castles, Ireland's impressive ruins ignite romantic reflections in many. Yet, just like the modern ruins of ghost estates, they also tell of an often oppressive relationship between human cultures and the natural environment. Ironically perhaps, stone circles and tombs that seem to speak of people living in much closer relation to non-human nature than we moderns do are also associated with the environmental scourge of deforestation. Yet,*

Environmental Philosophy: The Art of Life in a World of Limits
Advances in Sustainability and Environmental Justice, Volume 13, 175–195

ISSN: 2051-5030/doi:10.1108/S2051-5030(2013)0000013012

they at least stand testament to an ethic of timelessness and robust building, as well as resistance to a seemingly irresistible process of capitalistic modernisation; the recent ruins are devoid of such ethical commitments. Given this, however, creative responses should also be noted to the logic of the ghost estates, including Cloughjordan's Ecovillage and the NamaLab project.

Practical and social implications – *Three sets of responses that all work more realistically with a recognition of the limits of sustainable development are considered in the conclusion: Transition Towns, an Ecovillage and architectural reutilisation of defunct buildings.*

Keywords: Ruins; Ireland; sustainable development; environmental ethics; cultural relations with nature

INTRODUCTION

Ireland, as we all know, is dotted with ruins. Crumbling old walls overgrown with ivy, where the caw of rooks overhead adds plaintively to the romance of standing amidst time that is long past but still alive to the imagination. Linear stone ledges, their foundations reaching out of the rocky ground from which they were hewn, are broken open by trees reaching towards light where once a proud roof held the elements at bay. Whether the humble remnants of a disused house, the enigmatic megalithic remains of a long lost people, or the stone skeleton of a more recent but still ancient religious site, we marvel at these reminders that some things were built to last. The ruins of Ireland present a concrete link with the deep history of far western Europe and its people, especially when compared to Australia, where a 150-year-old wooden hut is considered an ancient relic of England's pioneering days. While the history of British colonisation remains a serious political issue in both countries, it is the paradigm of unstoppable technological progress and ideologies of the never-ending growth of capital that continues to devastate the social and environmental health of both lands. This pattern represents the modern extension of an ancient model, which strives to exercise cultural 'mastery' over the environment as well as centralised dominance over the general populace. Such a pattern is nowhere more visible than in Ireland's new ruins: the 'ghost estates' as they have become known.[1] Examining them in the context of the history of ruins provides rich insight into a deep pattern of colonisation, which marks late modern housing bubbles, but also shows genealogical roots in the long history of civilisation and its built environment.

Ireland's ghost estates are instant ruins that symbolise the recent history of capitalist overdevelopment. They are also symptomatic of a wider malaise that marks the very ideological heart of civilisation itself, from its initial premises to the contemporary utopian dream of global society. This is the dream that we may all live like royalty in a city of abundance and never have to pay the full price of our appetites (Berry, 2013). This unsustainable utopian dream is defended by all major economies that remain devoted to growth even as they peddle out a conciliatory dish of pale green rhetoric. The dream is cloaked in rational action even while it perpetuates an entirely irrational habit of overconsumption, while it is sponsored by a mass media myth that everything will work out as long as we keep consuming. At exactly the same time it sells sensational stories about the extreme weather results of climate change that offer no creative response at all to its real dangers.

With so much of humanity locked into the habits of overdevelopment that are the hallmark of today's dominant modes of production and consumption, there comes a disturbing disregard not only for the disadvantaged members of society but for the earth itself, as a web of living entities with intrinsic rights and values beyond our instrumental purposes and desires. This disregard is not necessarily limited to large-scale settlement societies, but such a way of life exacerbates the human tendency towards anthropocentrically focussed environmental destruction. Comparison between the cultural specificities of ancient and new ruins reveal both a wide disparity between building styles and aims – the shift from sturdy stone to flimsy prefab the most immediately obvious – and an ongoing tendency towards human domination over the land that infects the project of civilisation from its inception.

IRISH RUINS ANCIENT AND NEW: MEGALITHIC ROMANCE AND ENIGMA

Sometimes restored to touristic magnificence, ruins dot the Irish town and countryside so ubiquitously that they may be found, as was the example shown in Fig. 1, unmarked at the end of an unremarkable cul de sac. Like many visitors to Ireland, I cannot resist the lure of these ruins, but also regard them with mixed feelings; respect for the skill of their constructions,

Fig. 1. Unmarked Ruin, Co. Limerick (Photo by Author).

but also with a pang of recognition in the realisation that all things must pass; with honour at the durability of formerly proud holdings, but also sometimes a feeling of shame at violent overlords and their oppression of those 'beneath' them. The ruin in Ireland, like everywhere, often represents a dominant perspective over the surrounding landscape and everything upon it. In this particular slowly crumbling castle, shaft-like window slots remain, once providing cover for defensive bowmen, who enjoyed the elevated vantage point provided by the lords of this powerfully built edifice. But the ruin can also signify intimate relations with the land and its other creatures, as it does in the stone circle or tomb.

Sacred sites such as those shown in Figs. 2 and 3 represent a profound connection between people and the world they inhabit. The mound is often smoothed off in a beautiful, egg-shaped ellipse, with impressive megaliths around the circle and a low tunnel-like burrow in the centre. It is a womb

Fig. 2. Mounds at Knowth, Co. Meath (Photo by Author).

Fig. 3. Stone Circle and Tomb at Loughcrew, Co. Meath (Photo by Author).

upon the earth, protruding out from ground level in the fullness of pregnancy, its alignment with astronomical moments of significance like solstices and equinoxes speaking of a profound awareness of the seasonal cycles that bring more light or darkness to our world, and therefore signal a recycling of matter, a promise of new life or a return to the hardship of winter. The tomb is an inner sanctum celebrating the mystery of the night, a physical passage that is also a corridor between worlds old and new, deceased and refreshed. There is a clear connection to the sacred feminine at such sites, to the honour of giving birth and of being born, the life giving matrix out of which we all appear and back to which we all return.

To enter a stone tomb is to find silence, a place where contemplation is natural, where one can reflect upon life, mind, body and soul and find a way towards whatever understanding of or place in the universe one can achieve. There's no pulpit, although carved stone symbols often speak the ancient language of myth and vision. The bone fragments found here by archaeologists are not necessarily the remains of great leaders but may have been taken from the recent dead at a time of ritual, as was pointed out to me by Michael Quinley, a local historian and archaeologist who spent some time with me at the Grange stone circle, Lough Gur, Co. Limerick. Fresh batches of bones may have been used symbolically, as a way to direct attention back towards the death that awaits us, as well as back towards the ones that gave us life, ancestors primeval and recent. Inside such spaces, when enclosed, the air is moist but not close, earthy but not oppressive. The space is dark but embracing, not threatening. The shape invites you in, the low roof and entrance a humbling reminder of the human place in the universe. You may enter and leave alone but you share the place with others – past or present – and you are reminded that we must all make space for each other while we are here. Sometimes there is a pillar in the centre, which roots all space to the centre of the earth and our bodies. Sometimes there is a spiral on the wall, as we all dance around the centre of this life in cycles but never return the same. Everyone leaves something behind in the tomb but it never fills up. And we all get the chance to experience life afresh after each visit.

Entrancing spiral carvings (Fig. 4) can remind us that a sense of embeddedness in the natural world is often aligned with a cosmic connection. But this alluring combination shouldn't erase our understanding of the physical realities attending such constructions, or the wave after wave of colonisation that was involved in their creation. Before the agriculturalists, who had the technology to build these sites, hunter/gatherers roamed the sacred green land of Eire and left the forests standing. The megalithic builders were also farmers responsible for a significant amount of

Fig. 4. Spiral Carvings on Kerb Stone, Knowth, Co. Meath (Photo by Author).

deforestation and there is evidence of considerable soil erosion in places like the Burren that attests to the damage that result from such practices (O'Connell & Molloy, 2001, Dec; Drew, 1983, Feb). It may not occur to the extent that it does today, where entire hillsides are razed for profit; but even Neolithic deforestation was severe enough to contribute to massive loss of topsoil. This is a strange coincidence given that the stone circle reveals a style of building that is aimed to last, to forge a link between those living and the ancestors, but also between them and future generations, even those here millennia later. The significance of sacred sites built in stone speaks to such descendants and it says 'build as if it matters'. Because if you do it right, your own descendants will look back and be thankful that you saw through the ages and knew that just as our forebears passed life on to us, we pass it on to those in the future. And that is where the comparison with the ghost estates is so marked, because the new ruins were built for short-term profit in an age gone mad with greed and selfishness, standing against every conceivable principle of lasting quality and sustainable relations both between different levels of modern society and between human cultures and the rest of the earth. I want to begin to deepen my comparison between different eras – and orders – of building by first discussing the similarities between them. To do so necessitates leaving the tantalising world of megalithic sites behind and returning to the comparatively recent ruins of castles and keeps, cathedrals and ring forts.

The Staigue Fort (Fig. 5) is a proud construction that has survived nearly 2,000 years on the Kerry Peninsula. The fortress mentality reminds us of the landholder's interest in building to keep out those who would want to raid the fort for the wealth it stores behind its walls; the 'invaders' from without. Historically, the walls also represent a significant advantage when inclement weather strikes – and in this climate, that would have been often. Wind at body level is kept out, while solid walls give roofing a substantial base. Many ring forts are found in 'naturally defensible positions' (Hughes, 1972, p. 20), uniting one population in its defences against other populations as well as against challenges of the climate. It is in this sense that David Lloyd uses the term 'Fortress Europe' to consider the modern conglomeration of states brought together under the united currency and other arrangements of the European Union. This fortress is designed to keep other people at bay – this time preventing potential third-world invaders from entering economically developed nations, or keeping welfare recipients in nearby but socioeconomically disadvantaged suburbs away from the wealthy, or simply marginalising the rural poor made redundant by the

Fig. 5. Staigue Fort, near Sneem and Caherdaniel, Co. Kerry (Photo by Author).

industrial mechanisms that helped to build the pyramid of profit in the first place. But even at the smaller scale of the ghost estates, we see the ring fort mentality of the protected enclave take root, with severe repercussions in terms of social justice (as the people pay for the faults of the bankers) and ecological destruction.

RUINS AS A METAPHOR FOR RESISTANCE TO THE MODERNISATION PROCESS

While the ghost estates are Ireland's new ruins, there is nothing majestic about them, as there is in the crumbling old castle or keep. Even the romanticism of decrepitude is undone by the realities of 'austerity' and the hardships of the poor. It is easy to be swept away by broken walls or stone alignments covered in ivy on a windswept hill, but we're not so enamoured of empty houses that were never built properly in the first place, while homeless people shiver through another frosty night. The place of Ireland in this is uniquely interesting because of the traditional Irish resistance to urbanisation (Evans, 1996, p. 32), as well as because of the jarring disparity between the highly touted and touristed romance of ancient ruins and the shameful existence of recently constructed ghost estates. In many ways, we end up with modern ghost estates for similar reasons that Ireland is dotted with ancient ruins: because we all want to live like kings. The old castles took advantage of elevated positions for commanding views and the associated benefits of surveillance and security; the ruler lorded it over land and lesser mortals alike, while kin and servants benefitted (and/or suffered) from proximity to power and the abundance it proffers.

The old churches likewise often preferred the hilltops, which gave them the advantage of being closer to God and further away from the lowliness of nature. The medieval religious perspective desired a view from the heights, ostensibly to position its sacred space as close as possible to the God on high. Doubtless this position is often filled with sincerity, but just as surely it can be seen as part of a long history of assuming cultural authority over people and land. This can be seen when churches are built over pagan sites of ritual or consecration, as well as when they work hand in glove with a monarchy or other form of ruling elite. Either way, this remains a situation of power – the view over the people and the land represents a dominion enjoyed from above, the call to the faithful who look up to the exalted ones, cloistered or otherwise; a conduit to their highest hopes and dreams (Figs. 6 and 7).

Fig. 6. The Rock of Cashel on High, Co. Tipperary (Photo by Author).

Fig. 7. The View from Cashel (Photo by Author).

But Ireland's ancient ruins aren't all about domination of the land and the people. As we saw with the megalithic remains, they also speak of a time when technology sought alignment with the laws of nature – of life and death – even while this also contributed to deforestation on a sometimes drastic scale. Then there is also the humble homestead, the ruin that represents a kin group and their attempt to enjoy shelter, who may just as well have improved the land upon which they dwelt (Fig. 8).

Thought about symbolically, ruins can also be considered a metaphor for the survival of that which resists exploitation, according to David Lloyd. This is because they are judged as premodern, a residue of ways of life that the regime of profit needs to eradicate so that it can roll over the land unhindered. Ruins, like the myths and lifeways of those who continue to identify instead with the rest of nature, are the remnants of a past that stands between Ireland and its full membership in 'Fortress Europe', with its calamitous 'superstate of capitalist domination', which 'continues to widen the gap between the world's rich and the world's poor' (Lloyd, 2008, p. 2). While Lloyd takes a rosier view of ruins than I, he is not claiming that these remnants are 'regressive images of some impossible Golden Age' – rather, they remind us of alternative ways of living, stubbornly

Fig. 8. A Humble Homestead Somewhere in Central Ireland (Photo by Author).

resisting the way the past is relegated to a lesser place (p. 3). Ruins are like myth, 'open to both the past and the present', usually forgotten by the machinery of modernisation while conserving 'the outlines of utopian desires that might challenge the dogmas of modernity' (pp. 3–4). Defending against the 'work of modernity' – which for Lloyd is 'to obliterate both the memory and the present consciousness of its violence, and to naturalise progress as the self-evident form of human time' – the ruin 'both recalls destruction and comes into conjunction with the obstinate refusal in the present to accept that there are no alternatives' (p. 4).

While Lloyd's work is designed to serve the interests of social justice, his stand against 'domination in any form' also makes him an ally of the move towards greater ecological accountability in Ireland, as it opposes the state as a 'conduit that channels the interests of corporate capital' and thereby subsumes 'every means and mode of life to the processes of production and profit' (pp. 8–9). Lloyd is not just thinking of the majestic ruins rightly touted as some of Ireland's most prominent tourist destinations; he is also thinking of the share-farmers' humble homes, left deserted when the process of modernisation has made their lifestyle impossible to retain. (Ake Campbell had noted of the Irish rural house in 1937 that it 'never stands out in bold relief against its background but melts into it even as a tree or rock'; Evans, 1996, p. 63.) Lloyd's commentary on the economic and administrative modernisation process from the mid-19th to late-20th centuries leads directly towards critique of the ghost estates. The earlier development proceeds according to the logic of clearing and enclosing the land and evicting the populace (p. 13), destroying local populations that still shared 'certain forms of common ownership and economic reciprocity ... to allow for the concentration of land and the extraction of surplus value to take place' (p. 14). Lloyd traces out the way the modernisation process encloses space according to the English model of orderly geometry of land over the seeming confusion of Irish (communal) space. This process parallels Michel Serres' recognition that the straight lines and 'just measure' of commerce ignore the meandering routes of the human adventure and the possibility that self-understanding could be more valuable than profit (Serres, 1982, pp. 52–53).

The modernising mentality dismisses the ruin just as it dismisses myth; both represent a 'dimension of loss' that 'should have been developed out of culture by reason' (pp. 15–16). Everything, as Adorno and Horkheimer noted in their *Dialectic of Enlightenment*, is commodified by the relentlessness of this logic. But, as Lloyd reminds us, the spectres of ruins and hopes for a more just world, both socially and environmentally, haunt modernity as the ghosts that it created (pp. 39–43). The idea that ruins should survive to haunt late modernity would come as no surprise to geographer E. Estyn Evans, for whom Ireland's 'far western heritage, poor in material resources but rich in spiritual values' always resisted change and worshipped 'other things than progress' (Evans, 1996, p. 67). Here, Lloyd's identification between myth and ruins, which jointly resist the steamroller of modernisation and its profit motive, becomes even more poignant. The loss of Irish mythology as a living force in everyday life removes the protective force of the female figures traditionally used to personify the country, as 'wishful symbols of the fertility of the land and the people' (Evans, 1996, p. 68). Following Flower's portrayal of the Dindshenchas (or lore of places and place-names, Flower, 1947), Evans laments the passing of 'a sense of the harmony and mystery of man's place in nature', when the 'gods dwelt among the hills, and the living spirits of the land were ever present' (1996, p. 68).

It is too easy to simply dismiss such observations as mere romantic yearning after a lost past; as Lloyd pointed out, they also represent alternatives to the seemingly unstoppable dominance, injustice and ecological devastation of global capital. The perseverance of Irish myth is closely associated with 'the roots of such marked Irish traits as an addiction to story-telling', which Evans suggests are analogous with the ancient traditions of hunters and fishermen in Ireland and other premodern 'folkways of food-gathering societies surviving to this day in other parts of the world' (Evans, 1996, p. 71). Thus Irish roots 'deep in the Mesolithic substratum, however much they have been altered and added to' (p. 71) bring us back to the Australian Aboriginal traditions, which represent the oldest continuous cultures on the planet. The renowned conservatism of the Irish farmer bears similarities with that of the original Australians, who also represent on the whole 'conserving societies' that fit cultural patterns to a particular environment in order to survive its challenges (p. 71). The original Australians identify with the land, or their 'country', to a level almost impossible to understand to the modern urbanite. The fact that the native Australians built very rarely and never to the extent of the megalithic peoples of course marks an enormous difference between them and ancient Irish populations.

The difference is probably found in the history of agriculture, which swept across the British Isles into Ireland well before written records began: by the end of the fourth millennium BCE Irish 'farming communities were sufficiently stable to erect in the forest clearings elaborate stone monuments to serve as centres of burial and ritual' (p. 72). But the link between the two cultures also remains, because like the Australian Aboriginals the builders of the megaliths were fixated on the place of humanity both in the rounds of the seasons and in the wider cosmos. Despite the enormous differences, then, indigenous Australian and early agricultural Irish societies share a propensity to define themselves in relation to the material world around them that is resistant to the monolithic profit motive of modernity, even when the megalith builders took up technological developments (such as the capacity for deforestation) that proved sometimes self-defeating. Ireland's megalithic builders therefore represent part of the core of the western European heritage (Fleure, 1934 in Evans, 1996, pp. 75–76) in a way that remains meaningful in terms of mounting a challenge to the devastating impacts (upon the poor and the environment) of global modernity and its capital.

The interval of Christianity and secular modernity between the traditional Irish and the modern era may have turned Irish thought 'away from nature' (pp. 68–69), but perhaps global modernity can find part of the inspiration it needs to return to better relations with the rest of nature by looking to the best of earlier and/or alternative traditions. Indigenous societies like those in pre-megalithic Ireland left no records, but we know from immediate observation that the indigenous Australians forged a viable way of life, often in marginal conditions, that lasted for at least forty millennia. And while the builders of the first ruins may have hastened the demise of some of their formerly pristine habitats (and cohabitants) with the deforestation that provided the building technologies for their monuments, they still sought to fit in with the natural order in some way that is no longer even considered by the builders of the ghost estates. The spiritual values that survive into modern Irish society, which can still often outweigh economic considerations (Evans, 1996, p. 86; and personal observation), provide a rich resource for the resistance to socially and ecologically devastating building such as we witness with the ghost estates. Interestingly, however, while Irish literature is deeply, movingly poetic and even speaks English back to the colonisers in a way that forces them to rethink their own language, it doesn't speak of the land much at all (Moss, 2000). Ireland may still be the spiritual centre of Europe (Moss, 2000, p. 10), inclusive and mythic (p. 12) and unremittingly poetic (pp. 29–30), but it treats the land as a backdrop to social action, which is strange in a culture renowned for its earthiness. Perhaps this indicates part of the reason why centralised political machinations have resulted in the dissolution of peoples' traditional links to the land. For as Luke Gibbons has noted, 'the state seems willing to leave cultural policy to unbridled market forces, as if the megastore is the custodian of a megalithic past' (Gibbons, 1996, p. 81).

GHOST ESTATES AND THE ECOLOGICAL CRISIS: MODERNITY, DISREGARD FOR THE REST OF NATURE, AND PROFIT

The ghost estate may not enjoy the same physically elevated or naturally venerated sites as the ruins so often do, but the attitude of domination over the land that is seen in the high castle or proud cathedral reveals a similar pattern of biases. The modern developer, financed by almost invisible funding corporations, looks over the land, with an eye towards profiting from it. He considers other people, the consumers to whom he markets the dream – of new homes, with neat yards and freshly painted walls and brand new washing machines – as numbers in his soon-to-swell bank account. The customers, in turn, buy into the dream, purchase their part of the aspirations shared by so many of us – to live like royalty – and thereby help to shift profit up the pyramid to the overlord. And while the developer climbs another couple of notches on the social scale, the banks are ultimately the biggest winners, virtually untouchable even when they fail. While the ghost estates and other disastrous housing developments are funded, the land and its other creatures will simply be cemented over; more lost habitat, more polluted streams, less arable land, less insects on trees for birds and foraging grounds for other mammals. Even worse, the new but often empty Irish housing developments will probably be ecologically disastrous in the long run even if they become populated, because they promote car-dependent families with no recourse to localised social infrastructure like health and education.

And the benefits of surveillance, never lost on the builder of the castle, with its slotted windows and elevated turrets, also come in new forms. The new overlord – the developer – can protect his client vassals (while they can afford the mortgage for the dream) with 24/7 cctv or permanently armed security alarms. Like archers at the stone window slots, the cameras and alarms operate to keep the outsiders outside, to give the almost gated community that feeling of security they would have had when they saw the king's men line up, well-armed and ready to fire on raider or starving thief alike. Those who profit the most get to protect their investment with the best arms, whether that be crossbows or video cameras backed up by the police and the law. Someone at the top always enjoys the best pickings and someone below – servant class and peasant both near and far, as well as otter and songbird and beast of burden – always struggles more just to survive. The empty ghost estates reveal the shadow of this play of domination just as the crumbling ruins do. And while the people suffer unnecessarily under the yoke of cruelly selfish leadership, so too does the land, which is denuded by the pressure to produce more in lean times.

The ruins of ghost estates represent the worst of architecture and social and environmental planning (Fig. 9). They are ruinous to biodiversity, destroy arable land, create and sustain unproductive monocultures and are marketed with recourse to a utopian dream that is concurrently a dystopian nightmare. These instant ruins perpetuate the ancient dream of civilisation's mastery over the rest of the earth, an attitude that is inherently exploitative of people, land and other creatures. But the ghost estates, as Luke Gibbons points out, represent not so much 'a clash between nature and culture [as] between several opposing cultures, the debris of a history of invasions' (Gibbons, 1996, p. 159). This is very different to the tragic vision of the romantic aesthetic, according to which, as Gibbons put it, ruins came to represent 'the triumph of natural forces over human endeavour' (p. 159). While the 18th century afforded initial insight into the damaging tendencies of industry, it also allowed that such machinations

Fig. 9. An Irish 'Ghost Estate' circa 2012 (Photo by Author).

would inevitably give way, as everything does, to the sands of time (think Shelley's 'Ozymandias'). Little could the Romantics have known that not only would capital overcome all obstacles in its path of profiteering mania, but that it would even overwhelm the best of climate science and the global financial crisis of 2008 to continue its march towards further ecological collapse. The ghost estates may look like a symbol of failure, but they actually represent the 21st-century triumph of capital, in this case mobilised by one segment of Irish society to the cost of all the others, in the name of ongoing, unethical and environmentally disastrous profits.

It is in this sense that Anthony Haughey's photographic images of ghost estates and half-built houses exhibited at the Copper House Gallery in Dublin ('Settlement', Haughey, 2011) recorded 'not a past to be mourned in passive melancholy' but 'a political and social outrage, a product of bad economics, bad planning and bad politics' (O'Toole, 2011). 'Ghost estates', wrote O'Toole,

> are not the relics of glory. They speak to us not of something that used to be but of something that never was. Not only were they abandoned before they were inhabited, but the thing of which they speak – the Irish bubble – was an illusion. They are the vestiges not of a great empire but of a *folie de grandeur*.

Avoiding mere 'aestheticisation of what is, after all, living misery' – an aestheticisation almost inevitable in the photographic medium, as Susan Sontag noted long ago (Sontag, 1978) – Haughey 'consciously shaped Settlement as a positive political response to the world of negative equity, ghost estates and Nama' (O'Toole, 2011).[2] Haughey carried on an earlier interest in environmental and political disruptions to the Irish landscape, joining other commentators in suggesting hope for the future by 'making something beautiful' out of 'the tiny amount of available light' in the prevailing darkness of the ghost estates (O'Toole, 2011). Haughey's exhibition was also linked to the NamaLab project carried out by students and staff of Dublin Institute of Technology, who set out proposals for the creative transformation of empty buildings.

Other photographers to record Ireland's ghost estates include Kim Haughton, whose 'Shadowlands' images (Haughton, 2010) include some of the most aesthetically and politically charged photographs imaginable on the topic, and Valérie Anex. Anex's shots reveal the ruined landscape, the monocultural death of biodiversity and even the dream used to advertise the buildings, with one picture featuring a real estate marketing poster forlornly taped up to a dead grey cement brick wall (MacDonald, 2011, image #3: 'Battery Court, Longford.'). Anex's stated aim is to turn attention to a very simple question: How do things like this happen? In attempting to answer this question, she talked to people and found anger in their responses. As Kerri MacDonald reported, the people Anex met while photographing 'felt as if the character of their communities has been destroyed. Nobody, she said, wants a ghost for a neighbour' (MacDonald, 2011). This is obviously a very different response to the survival of ancient ruins, which represent a tourist bonanza as well as an understandable element of pride in the dead and long uninhabited.

But not everybody retains such anger. While it has become apparent that 'everybody ultimately lost in the property frenzy', Rob Kitchin writes that most people are also 'getting on with their lives and whilst resentful of the haunted landscape of excess property still love the place they call home' (Kitchin, 2012, 'Deserted Village'). There is even hope in the fact that 'some unfinished estates are now starting to be levelled and returned to greenfield status' (Kitchin, 2011, 'Demolishing unfinished estates'). Kitchin asks why such situations can't more easily be resolved with constructive input by developer, bank and council, especially when sites have been described as dangerous (posing health and safety concerns due to large holes and absence of secure fencing). He and Denis Linehan position themselves against what they see as the dystopianism of David McWilliams, who 'bleakly promoted a common idea now in circulation that the Ghost Estates are ghettos-in-waiting' (Linehan, 2009). McWilliams had been comparing blighted neighbourhoods of US cities and the future of Irish housing, pointing out that empty homes wound the national psyche and act as loci for anxiety. Linehan, however, considers these empty housing estates as 'better understood as part of the peculiar geography of the *NAMAscape*, whose topography still needs careful mapping and interpretation': a problem best tackled not with bleak dystopian visions but with 'creative, imaginative and innovative housing policy which is still, despite the crisis in government finances, not outside our grasp' (Linehan, 2009).

Resolving the social and environmental issues surrounding the ghost estates requires a significant process of retrofitting – something that is often associated with the idea of Transition Towns, which share such optimism in an oil-reduced and climate challenged future (and which grew from the experiments of Rob Hopkins in Kinsale, Co. Cork). An even more forward thinking project found on Irish soil is the Ecovillage in Cloughjordan, Co. Tipperary. While the photograph in Fig. 10 may share some aesthetic similarities with ghost estates – some buildings and grounds are unfinished, for instance – the consciousness could not be more different. Where the unscrupulous developers of ghost estates threw up poorly designed houses at a furious rate of knots, throwing caution to the wind and then

Fig. 10. Cloughjordan Ecovillage, Co. Tipperary (Photo by Author).

abandoning ship when the tide turned against them and their easy credit, the planning behind Cloughjordan was meticulous and careful. Founding members began formulating the project in 1999; by 2003 much of the capital was raised and the ethos of shared responsibility agreed upon; land was purchased in 2005 (after community consultation); and building was begun in 2007 (infrastructure) and 2009 (homes; see Cloughjordan 'Background history'). This is certainly a case where time has been taken to ensure quality of living, with a palpable sense of community and a manifesto designed to incorporate environmentally aware principles (such as building to PassivHaus standards, using energy efficient lime-hemp insulation and biodegradable cob materials, and many recycled products such as rubber for roof tiles and reclaimed timber for a wide variety of other uses; see Cloughjordan 'Construction Case Studies').

The Ecovillage project at Cloughjordan combines the best of the positive principles I have found in Ireland's ruins – it is built to last, provides a sense of community and aligns human culture with the interests of a broader nature, which includes principles of sustainable building and energy efficiency. Of course most of us live in places where this level of planning never took place, which leaves us with the type of creative responses found in Transition Towns as a path towards improving ecological standards of living. The other option is to ignore the clarion call for change that is found in climate science, to allow corporate raiders to continue exploiting the earth's precious resources for profit while doing untold damage to the environment, and to have faith that technology will continue to evolve quickly enough to save us (and the planet) from ourselves. But the history of Irish ruins reminds us that such faith may be ill-placed. The builders of megaliths had greater technology than the earlier Irish hunter/gatherers, yet themselve became a distant and enigmatic memory now lost in the mists of time, leaving behind their stone buildings but no other records. Iron Age ring forts gave way to another society, while castles held the high ground for many centuries, as did the cathedrals. All of these buildings became either tourist sites or crumbling ruins in the face of modernity and its more developed technologies and unremitting process of exerting mastery over the earth. The ghost estates represent this force at its worst – empty of community or soul, dismissive of the environment, wasteful and careless and in ruins before even habited. This is the kind of 'mastery' that threatens its own foundations, hopefully leading towards more thoughtful building and ideals of life quality, but certainly leading to embarrassing instant ruins with no redeeming features whatsoever.

NOTES

1. The term 'ghost estate' was first used by David McWilliams as early as 2006 and was soon taken up by the media and in everyday discourse (Kitchin, O'Callaghan, & Gleeson, 2012, p. 5). It is generally accepted to be an estate of 10 or more housing units where 50 percent or more of units are either vacant or under-construction (Kitchin et al., p. 5).

2. 'The formation of NAMA was announced in the Minister for Finance's Supplementary Budget on 7 April 2009, with the National Asset Management Agency Bill (2009) published September 10th of that year. The Bill enabled NAMA to acquire bank assets from five Irish banks relating to land and development loans and associated loans, and to manage those assets for the benefit of the taxpayer. The idea behind NAMA was to relieve Irish banks of their impaired assets, providing them with government-backed bonds which they could use to borrow from the European Central Bank, and thus inject liquidity into the Irish banking system. It would also have the effect of protecting both the banks and developers from going bust' (Kitchin, O'Callaghan, & Gleeson, 2012, p. 11). However it failed spectacularly to protect people or land, employing those who created the problem in the first place as experts in its solution and concurrently blocking creative responses (p. 13).

REFERENCES

Berry, G. (2013). The symbolic quest behind today's cities of light – And its unintended ecological consequences. *Journal for the Study of Religion, Nature and Culture*, *7*(1), 7–26.

Cloughjordan Ecovillage. *Background history*. Retrieved from http://www.thevillage.ie/index.php?option = com_content&view = article&id = 78&Itemid = 17

Construction Case Studies. Retrieved from http://www.thevillage.ie/index.php?option = com_content&view = article&id = 85&Itemid = 17

Drew, D. P. (1983). Accelerated soil Erosion in a Karst area: The Burren, western Ireland. In *Journal of Hydrology*, *61*(1–3), 113–124.

Evans, E. E. (1996). *The personality of Ireland: Habitat, heritage and history*. Dublin: The Lilliput Press.

Fleure, H. J. (1934). Prehistoric elements in our Heritage. *Bulletin of the John Rylands Library*, *18*(2), 325–358.

Flower, R. (1947). *The Irish tradition*. Oxford: Clarendon Press.

Gibbons, L. (1996). *Transformations in Irish culture*. Cork Ireland: Cork University Press.

Haughey, A. (2011). *'Settlement'* photographic exhibition, viewed at Copper House Gallery, Dublin.

Haughton, K. (2010). *'Shadowlands' gallery*. Retrieved from http://kimhaughton.photoshelter.com/gallery/Shadowlands/G0000ChCUrmHIMoY/

Hughes, K. (1972). *Early Christian Ireland: Introduction to the sources*. London: The Sources of History Limited.

Kitchin, R. (2011). Demolishing unfinished estates. Ireland After Nama blog, August 16 2011. Retrieved from http://irelandafternama.wordpress.com/2011/08/16/demolishing-unfinished-estates/

Kitchin, R. (2012). Deserted village. Ireland After Nama blog, March 24 2012. Retrieved from http://irelandafternama.wordpress.com/2012/03/24/deserted-village/

Kitchin, R., O'Callaghan, C., & Gleeson, J. (2012, February). *Unfinished estates in Post-Celtic Tiger Ireland*. In NIRSA (National Institute for Regional and Spatial Analysis). Working Paper Series No. 67, NUI Maynooth. Retrieved from http://www.nuim.ie/nirsa

Linehan, D. (2009). Ghost estates: It happened in the US, it could happen here? Ireland After Nama blog, Dec 5 2009. Retrieved from http://irelandafternama.wordpress.com/2009/12/05/ghost-estates-'it-happened-in-the-us-it-could-happen-here'/

Lloyd, D. (2008). *Irish times: Temporalities of modernity*. Dublin: Field Day Books.

MacDonald, K. (2011). *In Ireland, ghosts of towns that never were*. Retrieved from http://lens.blogs.nytimes.com/2011/08/22/in-ireland-ghosts-of-towns-that-never-were/

Moss, J. (2000). *Invisible among the Ruins: Field notes of a Canadian in Ireland*. Dublin: University College Dublin Press.

O'Connell, M., & Molloy, K. (2001, Dec). 'Farming and woodland dynamics in Ireland during the neolithic' biology and environment. *Proceedings of the Royal Irish Academy* (Vol. 101B, No. ½) *From palaeoecology to conservation: An interdisciplinary vision* (pp. 99–128).

O'Toole, F. (2011). Ireland in Ruins: The height of folly, the depths of misery. *The Irish Times*, Saturday, November 12, 2011.

Serres, M. (1982). *Hermes: Literature, science, philosophy*. Baltimore, MD: Johns Hopkins University Press.

Sontag, S. (1978). *On photography*. London: Allan Lane.

WALKING RESPECTFULLY ON THE EARTH: A PROBLEM FOR DEEP ECOLOGY

Thomas Duddy

ABSTRACT

Purpose – *To explore an idea that has been articulated and defended by a number of influential deep ecologists, namely, the idea that all living organisms or species should be valued equally, or regarded as possessing equal intrinsic value.*

Design/methodology/approach – *Examination of some of the key works of deep ecological theorists.*

Findings – *Despite its initial attractiveness as an expression of respect for all living things, there is something fundamentally paradoxical about the deep ecological position that has come to be known as biocentric egalitarianism.*

Research limitations/implications – *What is presented here is partial and does not cover all of the deep ecological thinking.*

Practical implications – *Offers some initial lines of exploration for scholars interested in the problems with this core deep ecological notion of biocentric egalitarianism/respect for all living things.*

Environmental Philosophy: The Art of Life in a World of Limits
Advances in Sustainability and Environmental Justice, Volume 13, 197–210

ISSN: 2051-5030/doi:10.1108/S2051-5030(2013)0000013013

Originality/value – *Offers a new account of non-egalitarian biocentrism*

Keywords: Deep ecology; biocentrism; biocentric egalitarianism; intrinsic value

In this chapter I am going to focus on an idea that has been articulated and defended by a number of influential deep ecologists, namely, the idea that all living organisms or species should be valued equally, or regarded as possessing equal intrinsic value. Biocentric egalitarianism is defended explicitly by Paul Taylor who writes: 'Human beings are members of Earth's Community of Life in the same sense in which other living things are members of that Community', later adding that human beings 'are not inherently superior to other living things' (1986, pp. 99–100). Even philosophers who are not as univocally committed to egalitarianism as Taylor sometimes gravitate towards a position that is not very far removed from it, a position that seeks to radically reduce the difference between the value of human life and the value of non-human forms of life. Arne Naess, for example talks in terms of all living things having, in principle, 'the equal right to live and blossom' (1990, pp. 164–165), while Bill Devall interprets and defends biocentrism as the belief that 'all organisms and entities in the ecosphere … are equal in intrinsic worth' (Devall & Sessions, 1985, p. 67). Lawrence E. Johnson, employing a more holistic vocabulary, argues that we ought to seek to find our way in the world by giving 'due respect to the widely disparate interests of other beings', adding, significantly, that this 'is not essentially different from what we ought to do concerning other humans' (1991, p. 278). Most recently, James P. Sterba has developed a set of environmental principles that are, he believes, 'clearly not biased in favor of the human species, and thus provide a defensible interpretation of the biocentrist's commitment to the equality of species' (2011, p. 168).

My argument will be that, despite its initial attractiveness as an expression of respect for all living things, there is something fundamentally paradoxical about the deep ecological position that has come to be known as biocentric egalitarianism. I do not believe that it is possible to model our relationship with other living species on the relationship we have with each other as members of the same species. It is not because morality involves some degree of mutual recognition of rights and obligations, but because a biocentric environmental ethic would require us to relate to the natural world and to other species in ways that would make our own human lives practically unlivable. When I say 'practically unlivable', I mean unlivable in any recognizable human form, including the different forms that human life took, socially and technologically, among our pre-industrial, pre-modern ancestors. What is at issue here is not any technologically advanced or socially complex form of life, but human life in any form or mode that might be described as flourishing, as expressive of the variety of basic human needs and faculties. I have in mind here needs and faculties that have to do with our basic biological or animal natures, and with the exercise of our specifically human capacities for survival, including our original capacities for hunting, gathering and harvesting. It is precisely because of our biological, animal natures that we cannot respect (the rest of) nature in the way that we may respect each other. It is because of our animal natures that we are as dependent as we are on the earth and everything that lives on it and in it. We are, to use the term coined by B-movie extraterrestrials, earthlings. As earthlings, our relationship with the earth is manifold, and includes needs and abilities that we share fundamentally with many other earth-bound animals, including the ability to walk or otherwise get about on the living surface of the earth – on the earth's biomantle, to use the scientific terms.

Because we are so accustomed to walking, we sometimes fail to realize what a necessary part it plays in our relationship to the natural world in which we survive and sometimes flourish. Perhaps because we have invented faster ways of getting about, from skateboards and bicycles to cars, trains and planes, we fail to value walking, except for reasons of health and fitness. In any case, walking comes so naturally to us that we do not think of it as having moral or ethical implications in itself. It is something that we do almost thoughtlessly, and this thoughtlessness is more significant than we might think. When we refer to walking in any sort of ethical context, it is usually in a negative sense. It is significant that when we wish to state graphically the extent to which one person fails to respect another we say: He walked all over her (or, she walked all over him). To walk all over someone, in some figurative if not literal sense of the term, is to show great disrespect for her or him. Yet, this is what we must do to get about in the (non-human) world – walk on it, step on it, otherwise travel on it, as we go energetically about the business of living.

And that's just the beginning of our manifold relationship with the living earth, relationships that are frequently utilitarian, sometimes violently so, as far as some other species are concerned. We build on the living earth, we dig into it, plough it up, plant things in it, harvest things from it, deposit waste matter and waste materials in it, and use it in various other life-preserving ways. We must continually destroy and consume things that grow out of the earth if we are to preserve and prolong our lives; having consumed or otherwise used things, we produce waste materials; having eaten we produce waste matter, which we must expel from our bodies and this continual production of waste materials and waste matter will always cause some degree of pollution, or potential for pollution. It is not possible for us to opt for zero-consumption, zero-waste, zero-pollution or zero-destruction (in the way that we can opt for zero-murder, zero-theft or zero-dishonesty). To put it another way, it is not possible for us to opt for absolute conservation or absolute preservation (in the way that we can opt for absolute respect for each other's lives, or for absolute honesty in our dealings with each other). We can define immoral acts such as murder, theft and lying as the kinds of acts that can be categorically or unconditionally prohibited – that is prohibited on principle, prohibited absolutely, without allowing for exceptions. It is possible in principle that a person could live a full and meaningful life without engaging in murder, theft or lying, or any other sort of immoral act. Immoral acts are acts that are always avoidable, as a matter of principle. *Ought* really does imply *can*. It would be impossible, however, to put the activities of consumption, pollution or local motion on a par, morally speaking, with the kinds of actions that are prohibited by our moral injunctions against murder, assault or lying, since some degree of consumption and pollution is necessary to life itself.

To put the point in more positive terms, it would be impossible to put the injunction 'Preserve all living things' on a par, morally speaking, with the injunction 'Love your neighbour as yourself', since a policy of absolute preservation would render our natural human lives impossible. Loving one's neighbour as oneself may be very difficult at times, but it is not impossible, and if everyone managed to do so all the time, then human life would be enhanced. Loving one's neighbour can be considered an unconditional good or virtue, but preserving all living things cannot be an unconditional good or virtue, since it would lead to the starvation and death of all those who would seek to practice such a principle. To opt for absolute conservation would mean never making use of any part of the natural world, or treating any part of it as a resource. Everything would have to be left as it is, not used, not worked on or consumed. Likewise with absolute preservation. To opt for absolute preservation would mean never consuming any member of any other species, whether animal or plant. Even walking about on the surface of the earth would become a source of perpetual moral anxiety, since many areas of the earth's surface are alive with small-scale and microscopic life, and so we frequently damage small-scale plant and animal life when we move around in the natural world, even when we do so carefully. For these sorts of reasons, we cannot entirely agree with Kant when he says that destructiveness is immoral – a comment he immediately qualifies by indicating that he has in mind the wanton destruction of inanimate natural objects.[1] Certainly, wanton destruction is morally lamentable, but destruction is not always wanton. The level of destructiveness that it takes to pull up a plant, break it apart and eat it – or perhaps to kill an animal and eat it – is a necessary level of destructiveness for a species that, like all other creaturely species, has to take its energy from organic sources outside itself. An attitude of respect for nature that would be on a par with our respect for ourselves and other members of our own species would make it impossible for us to live a natural life, since living implies the relatively destructive activities of hunting and gathering – or farming and harvesting – and consuming, and the relatively polluting activities of producing waste materials and waste matter. We are not pure spirits or disembodied moral agents, but first and foremost biological organisms that must live off the earth and the other species it sustains.

Once we unpack all the implications of what it is to live as a natural being, then we see that that we must regard some parts of nature as sources of food, energy and raw materials for clothing and shelter, and so on. We see that the sources of food are going to be things that are themselves alive, even if only as plants. We see that we must compete with other species for the earthly spaces in which to live, that we must sometimes treat other species primarily as means to our ends rather than as ends in themselves, that we must systematically give priority to our basic needs at the expense of the basic needs of some members of some other species. If we treated everything in nature primarily as an end in itself we could not proceed energetically and resourcefully with our own survival, or feel good about the survival of other living species, since all living species survive and flourish at the expense of some members of some other species, whether it be other animal species or species of plant life. There is a very serious limit to the degree of respect we can have for the living things that we eat and otherwise consume, for the grass-roots habitats and organisms that we walk on, for the biota that we displace as we stake our claim to the sites and spaces on which we build our own large-scale, heavy-duty habitats. All biological forms of life compete with, displace or use, other living things to ensure their own survival. This means that no biological form of life, no matter how conscientious it seeks to be, can value the species it depends on for basic survival in the same way that it values itself. To value something as much as oneself would at least mean not harming or destroying it in order to consume it. In the case of human beings, it is precisely because we are dependent on various species of organisms, large and small, for our existence that we cannot value all of them non-instrumentally, that is for their own sakes. Once I say 'I need to eat', or 'I need to sow and harvest' or 'I need to build a house here', I cannot next say 'I respect all living things, equally, for their own sake'. It is our very dependence on the existence of some other living things that relieves us of the obligation to value all life equally. It is this same dependence that deprives the most righteous among us of the opportunity to make a virtue out of the purported valuing of all life equally.

It is sometimes suggested by biocentric egalitarians that life is indivisible – that all life is equally valuable, that human life is not more valuable than the life of any other species. But it is misleading and, I believe, unscientific, to use the term 'life' so abstractly – so abstractly that specific differences between different species of living things are blurred beyond recognition. All living things are alive in specifically different ways, including ways that cause them to use some members of other species in self-interested, self-preserving ways that may be aggressive and destructive, as in the case of hunting and harvesting. It is not true to say that there are no objective differences between species. It may be true that the differences are relative rather than absolute, but relative differences are real and ultimately significant. The fact that we have evolved from earlier forms of life, and are dependent for our continued existence on the existence of other species, means that we cannot deny our dependence on other species. But the fact that we *have* evolved, and evolved to become a *different* species, within a framework of acknowledged origins and dependence, means that the difference is as real as can be and cannot be ignored. To remain human means to remain relatively different, living a different kind of life from other species and relating to other species in ways that do not blur or deny the difference. All talk of respect needs to be tempered by acknowledgement of these realities of specific difference. To deny specific difference by blurring boundaries is to fly in the face of evolutionary, biological realities.

It is also important to realize just how great the differences are between the more recently evolved species, including ourselves, and earlier forms of life, including pre-mammalian forms of life that make up other branches of the evolutionary tree. The differences to which I want to draw attention do not have to do with intelligence, sentience or any other feature that is sometimes appealed to in order to argue for the supposed superiority of humans and the higher animals. (I am not concerned with the question of superiority at all, since it is possible to talk about significant difference in this context without making claims about superiority.) The difference I have in mind has to do with the survival strategies of the countless species of small animals, insects and other minuscule organisms, including earthworms, flatworms, ants, centipedes, millipedes, caterpillars, weevils, beetles, aphids, bugs, termites, silverfish, grasshoppers, crickets, woodlice, spiders, mites, springtails, grubs and others creatures, that do not share the visible, well-illuminated sectors of the biosphere that are occupied by ourselves and other mammals – that survive in fact by keeping out of sight, sometimes using camouflage to ensure invisibility. This very large group of species includes those that biologists call 'soil biota'. These are species that live in the soil for all or part of their lives. The scientific textbooks will tell you that nowhere in nature are species so densely packed as they are in the soil. It has been said that there can be more living organisms in a handful of rich soil than there are people on the planet. The textbooks distinguish four classes of soil biota, consisting of animals or organisms that spend some or all of their lives in the soil, nourishing and enriching it in the process, ranging from megafauna (e.g. moles, rabbits, rodents), through macrofauna (e.g. beetles, earthworms, ants) and mesofauna (e.g. mites), to microfauna, whose size is measured in micrometres (e.g. bacteria, protozoa, roundworms).[2]

The fact that most species of soil biota exist in very large numbers beyond the limits of normal human perception may not seem very significant, but it becomes significant as soon as we begin to think in terms of how we might respect nature to a radically greater extent than we do. We tend to think of land animals in particular in a very narrow sense, as consisting of those animals that have a visible presence, that can visibly cross our paths at some distance from us. The more visible an animal is, the easier it is to avoid harming it. We can walk through a wood or a field or a garden and not harm any of the larger wild animals that we encounter there. Most of them will get out of the way in any case. But such animals are only a small fraction of the animal life in the wood, field or garden. Out of sight, hidden in the undergrowth, in the grass, under leaves, in the soil itself, are countless hard-to-see organisms and small animals that are such an integral part of the living earth on which we walk that it would be profoundly difficult to avoid disturbing or harming them, directly or indirectly, to some degree. If we undertook to really care about such creatures, even to the extent that we care about the larger animals, let alone the extent that we care about ourselves, we would be unable to move about on the earth with a clear conscience, and would be unable to take any joy or delight in our own survival and flourishing. It is arguable that we would take ourselves to the edge of sanity, unleashing a potentially paralysing sense of guilt about our very existence as a natural, large-bodied species of mammal. Of course, the impact of everyday walking on the earth is much less than the impact of some methods used in industrial farming and industrial construction, but even a relatively small impact would have to concern us if we were biocentric egalitarians who purport to value all life equally. 'All life' would have to include all species of small-scale life, including microscopic life.

These considerations give rise to serious doubts about the rationality and wisdom of biocentric egalitarianism. It does not seem good enough to argue, as Naess does, that we can be biocentric egalitarians in principle, if not generally in practice, since a principle that we cannot hope to realize generally in practice only serves to discredit the very mode of thought – in this case, ethically environmental thought – that has articulated and sought to advocate it. Rejecting biocentric egalitarianism, however, does not necessarily commit us to a narrowly or exclusively human-centred view of the relationship between the human species and the rest of nature. It certainly does not mean that we cannot cultivate an attitude of some significant degree of respect towards the rest of the natural world, especially towards species other than ourselves – and especially towards other animal species. Just because non-human species are not on a par, morally speaking, with human beings does not mean that they need not be valued at all, or that they should only be valued instrumentally. Why do we – or why do at least some of us – experience regret, even grief, when we learn that a species has become extinct or is on the verge of extinction? Some people – perhaps most people – are capable of feeling some dismay or regret when they hear reports of species extinction, including the extinction of plant species. The feelings of regret will not normally be as great as those that are more grievously felt when we learn of the deaths of other human beings, such as those that occur in large numbers as a result of natural disasters or wars, but there is some level of regretting just the same. This regret is felt independently of whether or not the extinct species has any direct relevance to human survival or flourishing. The regret may also be felt even when human activity has not been directly responsible for the extinction in question. It is as if there is something in us – or in many of us – that, other things being equal, causes us to experience a sense of loss when the earth's complement of life is diminished. It is as if the life in us, despite specific differences between ourselves and other species, responds after all to life elsewhere, preferring the continuation of such life, other things being equal, to the news of its destruction or extinction. If we must use the term 'respect', however, it should be in a suitably qualified sense, a sense qualified by our understanding of ourselves as animals who could not survive in a world of absolute conservation or absolute preservation.

I am going to offer some reflections on three poems that usefully address in different ways the feelings, intuitions and thoughts that are concerning us here. The first of these is Philip Larkin's poem, 'The Mower', which describes the feelings that pass through him and the thoughts that occur to him on finding that he has killed a hedgehog while mowing grass. He realizes that he has seen the animal before 'and even fed it, once'. He is shocked to have been the cause of the animal's death. Burying it has not alleviated his feeling of regret:

> Next morning I got up and it did not.
> The first day after a death, the new absence
> Is always the same; we should be careful
> Of each other, we should be kind
> While there is still time.
>
> (Larkin, 1988, p. 214)

The regret that Larkin feels in the aftermath of accidentally killing the hedgehog is praiseworthy and, I think, more natural than the absence of any such feeling. Yet, the idea that Larkin should regret this death to the same extent that he might regret the death of a fellow-human is not necessarily a good idea, nor a natural one. Larkin should earn our admiration if he were to take more care when mowing in future, so that he reduces the risk of killing another hedgehog. He might earn even more admiration by going further and deciding not to mow his lawn ever again, the better to reduce the risk of killing or harming anything that lives there. But such an act of goodness on his part, while morally praiseworthy, does not seem to be one that may be expressed as a moral imperative or duty.[3] We may find it regrettable that someone could cause the death of a hedgehog while mowing, and not regret doing so, but we would not think that an immoral act – an act of something like hedgehog murder – had been committed. While it is a moral imperative that we avoid doing anything that harms all innocent human beings, it is not so evidently a moral imperative that we avoid doing anything that harms all 'innocent' non-human beings.

This point becomes more urgent in the case of the second poem that I wish to reflect on. This is 'Moss-Gathering' by the American poet, Theodore Roethke, in which he describes his sensations and thoughts on having pulled up from the earth's surface a patch of dark-green moss, 'thick and cushiony, like an old-fashioned doormat', consisting of 'crumbling small hollow sticks on the underside mixed with roots, / And wintergreen berries and leaves still stuck to the top'. He finds that something 'always went out of him' whenever he dug up these 'carpets of green'. He always felt mean as he jogged back along the road:

> As if I had broken the natural order of things in that swampland;
> Disturbed some rhythm, old and of vast importance,
> By pulling off flesh from the living planet;
> As if I had committed, against the whole scheme of life, a desecration.
> (Roethke, 1885, p. 38)

The ecological sensitivity and sense of guilt expressed by Roethke is even more impressive, at first sight, than that expressed by Larkin. For one thing, we are dealing with botanical rather than just biological life. We are also dealing with a sensitivity that is more holistic in its scope, since the poet feels that he has offended against more than a living individual – he feels he has offended against the natural order, first of the swampland, and then of the living planet itself. He feels as if he has committed a desecration 'against the whole scheme of life'. The use of the word 'desecration' reinforces the sense that his offence is not only immoral but also unholy. The attitude expressed towards nature is closer to reverence than to mere respect. Is Roethke's sense of reverence even more praiseworthy than the more modest form of regret expressed by Larkin? I do not think that it is. To regard the earth, as Roethke seems to do, as something that stands over against us, as a vast object of reverence, something that cannot be touched, except in act of reverence, is to ignore the fundamental relationships of survival that exist between all living things and the natural environment in which they have to survive. The survival strategies of all animals involve energetic interactions, sometimes destructive, not only with the earth and some of its living species. This necessary energetic interaction is fundamentally instrumental, not reverential. An attitude of reverence is essentially a hands-off attitude, since to lay hands on the revered object is already to desecrate it. But this hands-off attitude is one that would threaten the survival of any natural being that undertook to practise it.

At this point, I want to consider a third poem, one that has a resonance with Roethke's, though it is from a different tradition of writing and thinking, namely, the Mexican Mayan oral tradition. This poem is really a form of ritual incantation, an incantatory song, that addresses in a graphic way the issue that we are reflecting on here. This poem – 'Before Felling a Tree', by the Mayan poet, Antonia Moshan Culej – is one of a number of such poems and songs that have been written down and translated by Ambar Past in her book, *Incantations: Songs, Spells and Images by Mayan Women* (2005). Two terms from the Mayan Tzotzil language are left untranslated, namely, Kajval and Kaxil. *Kajval* addresses the pantheon of supernatural protectors, including the sun and moon but also Cristo and the saints of Christianity, while *Kaxil* addresses the earth goddess. Here is the whole poem:

> Don't kill me, don't fall on me,
> Sacred Tree, Sacred Pine.
> It's because I am in need
> that I cut you down,
> Sacred Oak, Sacred Vine,
> Holy Earth,
> Holy Sky.
>
> Give me your firewood, your kindling,
> your torchlight so I can see what I'm eating, *Kajval*.
> Give me your heat to bake my tortillas, to boil my beans.
> Give me the beams to build my house
> and pillars to support the thatch, the vines, the mud.
>
> *Kaxil*, Godmother:
> I am going to split your wood, your arms, your legs,
> your face, your head.
> I'm going to chop you down with my ax,
> with my machete.
> Don't scold me, don't drip your tears on me.
> I don't want to cut you down, but it's cold.
> I want a fire to keep warm.
> I'm hungry, I have to make tortillas,
> I need light to see at night
> so I can grind my corn,
> so I can boil my hominy and beans.
>
> *Kajval*, give me your pitch pine.
> Your sunbeams for my candles.
> Thanks to you, I'm still alive.

The theme of this poem is the ambivalent and paradoxical one of trying to express reverence, while at the same time engaging in an act of necessary destruction. What is good about this poem is the sheer honesty of the admission to destructiveness:

> *Kaxil*, Godmother:
> I am going to split your wood, your arms, your legs,
> your face, your head.

> I'm going to chop you down with my ax,
> with my machete.
> Don't scold me, don't drip your tears on me.
> I don't want to cut you down, but it's cold.
> I want a fire to keep warm.
> I'm hungry, I have to make tortillas,
> I need light to see at night
> so I can grind my corn ….
>
> (Culej, 2005)

Some deep ecologists, including biocentric egalitarians, may find the ambivalence praiseworthy, but I see the ambivalence as revealing – revealing of the paradox at the heart of any religion or philosophy that preaches reverence or respect on the one hand and that accepts the natural need for survival on the other. There is a kind of fundamental, conflicted double-think at work in any attitude or philosophy that purports to revere the earth and everything on it, while at the same time insisting that a certain amount of sustained harm, destruction and displacement must take place in order for human life to continue. This double-think is understandable in the case of the Mayan poet who feels compelled to reconcile animistic nature-reverence with the need to survive, but it is not quite so understandable in the case of those of us who are not similarly committed.

For those of us who do not feel compelled to revere nature in the animistic way that the Mayan poet does, it behoves us, I think, to avoid the paradox that is revealed in the tree-felling poem. We need to face up – own up – to the fact that we are natural beings, that we live on the earth, that we live off the fruits and resources of the earth, and that there are therefore limits to the thoughtfulness with which we can move about on the earth's living surface. Because of our need to continually interact, sometimes destructively, with our natural environment, we do not normally feel regret at the destruction of *every* individual insect or wild plant. According to a Russian proverb, if you live near the cemetery, you cannot weep at every funeral. We might say in this context that, if we must move around energetically and resourcefully on the face of the earth, we cannot afford to weep for every individual insect or blade of grass that we step on. But we can justifiably feel regret on two accounts. We can feel regret at the destruction of whole species or life-sustaining features of the living planet. While our consciences with regard to other humans is focused on the individual person, our consciences with regard to nonhuman forms of life may be focused on ecological wholes – on species, ecosystems, wildernesses, the living earth itself. We could not have a fully natural human life without destroying some living things, whether animals or plants, especially plants, but we could live – and indeed *need* to live – a fully natural life without destroying whole species. More than that, our lives should be all the more fully natural for not subtracting from the sum of species and forms of life that constitute the biosphere. We cannot value all living specimens equally but we can value *species* to varying degrees, while at the same time refusing to regard any of them as valueless, other things being equal. Secondly, while it is true that we cannot be as mindful, as a rule, of the earth's small-scale fauna and flora as we are of ourselves, we can acknowledge our massive dependence on these grass-roots creatures by never willfully harming individual specimens when they come into the light or cross our path. This would mean taking a page from the book of the Jains who diligently avoid the unnecessary killing of small-scale, grass-roots life. We may not be able to live up to the Jainist practice of *Irya Samiti* or 'carefulness while walking' – which includes the practice of sweeping the ground where one is about to step – but we can avoid taking deliberate steps to kill the minuscule members of the animal kingdom that cross our path, and that are not a threat to us. The degree of respect-for-life expressed in avoiding such gratuitous killing or harming cannot be on a par with the kind of respect with which we ought to treat ourselves, but there is a degree of respect in it nevertheless. There is, in other words, a form of biocentrism, of respect for nature, that *is* defensible and adoptable, but it is a non-egalitarian form of it, not the ultimately unlivable, egalitarian form of it advocated by Arne Naess, Paul Taylor and others.

CONCLUSION

Yet, just as we may take a page from the book of the Jains, so we may take something with us from the work of the biocentric egalitarians. I see no reason why we cannot endorse to some significant degree the biocentrist principles recommended by Sterba, namely, (i) *a principle of human preservation*, which states that actions that are necessary for meeting basic human needs 'are permissible even when they require aggressing against the basic needs of individual animals and plants, or even whole species or ecosystems', (ii) *a principle of disproportionality*, which states that 'actions that meet the non-basic or luxury needs of humans are prohibited when they aggress against the basic needs of individual animals and plants, or even of whole species or ecosystems' and (iii) *a principle of human defence*, which permits actions that defend oneself or other humans against harmful aggression, 'even when they necessitate killing or harming individual animals or plants or even destroying whole species or ecosystems' (Sterba, 2011, pp.167–168). Indeed, it is arguable that the non-egalitarian version of biocentrism defended in this chapter is more consistent with these three principles than is the egalitarian version defended in a rather qualified form by Serba.

NOTES

1. 'Destructiveness is immoral; we ought not to destroy that which can still be put to some use. No man ought to mar the beauty of nature; for what he has no use for may still be of use to someone else. He need, of course, pay no heed to the thing itself, but he ought to consider his neighbour. Thus we see that all duties towards animals … and towards inanimate objects are aimed indirectly at our duties towards mankind' (Kant, 1963, p. 241).

2. For a readable, well-illustrated guide to the varieties of soil biota, see Nardi (2007).

3. Urmson distinguishes between acts that are good because they are in accordance with basic moral rules and imperatives and those that are good because they go beyond the call of imperative or duty: 'A line must be drawn between what we can expect and demand from others and what we can merely hope for and receive with gratitude when we get it; duty falls on one side of this line, and other acts with moral value on the other' (1958, p. 213).

REFERENCES

Culej, A. M. (2005). Before felling a tree. In A. Past (Ed.), *Incantations: Songs, spells and images by mayan women* (p. 136). El Paso, TX: Cinco Puntos Press.

Devall, B., & Sessions, G. (1985). *Deep ecology: Living as if nature mattered.* Salt Lake City, UT: Peregrine Smith.

Johnson, L. E. (1991). *A morally deep world.* Cambridge: Cambridge University Press.

Kant, I. (1963). *Lectures on ethics* (L. Infield, Trans.). Indianapolis, IN: Hackett.

Larkin, P. (1988). *Collected poems.* London: Faber & Faber.

Naess, A. (1990). *Ecology, community and lifestyle.* Cambridge University Press.

Nardi, J. B. (2007). *Life in the soil.* Chicago, IL: University of Chicago Press.

Roethke, T. (1885). *Collected poems.* London: Faber & Faber.

Sterba, J. P. (2011, June). Biocentrism defended. In *Ethics, Policy and Environment, 14*(2), 167–168.

Taylor, P. (1986). *Respect for nature.* Princeton, NJ: Princeton University Press.

Urmson, J. O. (1958). Saints and heroes. In A. I. Melden (Ed.), *Essays in moral philosophy* (pp. 198–216). Seattle: University of Washington Press.